Statistics for Scientists

Umberto Michelucci

Statistics for Scientists

A Concise Guide for Data-driven Research

Umberto Michelucci
Lucerne University of Applied Sciences
Dübendorf, Switzerland

ISBN 978-3-031-78146-9 ISBN 978-3-031-78147-6 (eBook)
https://doi.org/10.1007/978-3-031-78147-6

© The Editor(s) (if applicable) and The Author(s), under exclusive license to Springer Nature Switzerland AG 2025

This work is subject to copyright. All rights are solely and exclusively licensed by the Publisher, whether the whole or part of the material is concerned, specifically the rights of translation, reprinting, reuse of illustrations, recitation, broadcasting, reproduction on microfilms or in any other physical way, and transmission or information storage and retrieval, electronic adaptation, computer software, or by similar or dissimilar methodology now known or hereafter developed.
The use of general descriptive names, registered names, trademarks, service marks, etc. in this publication does not imply, even in the absence of a specific statement, that such names are exempt from the relevant protective laws and regulations and therefore free for general use.
The publisher, the authors and the editors are safe to assume that the advice and information in this book are believed to be true and accurate at the date of publication. Neither the publisher nor the authors or the editors give a warranty, expressed or implied, with respect to the material contained herein or for any errors or omissions that may have been made. The publisher remains neutral with regard to jurisdictional claims in published maps and institutional affiliations.

This Springer imprint is published by the registered company Springer Nature Switzerland AG
The registered company address is: Gewerbestrasse 11, 6330 Cham, Switzerland

If disposing of this product, please recycle the paper.

To Caterina and Francesca. To Francesca, the reason why all this is possible. To Caterina, my life, the person I am most proud of in my life.

Preface

This short book has originated from lectures I gave at various universities that have helped me identify the concepts in statistics that most scientists struggle with. Most of them have an intuitive understanding of what, for example, mean and median are, but few can tell when to use which, for example, or can explain how to interpret the coefficient R^2. This book is a summary of those concepts, described in a short and concise way, enriched with tips and rigorous explanations.

It is important to understand that the goal of this book is not to provide a complete course in statistics. There are many books that do this already. My goal is to give researchers and practitioners a short text that is easy to read to help them understand the most important concepts in statistics and especially how to use them properly. Readers are encouraged to continue the study of statistics, an incredibly interesting branch of mathematics that helps us understand the world around us.

This book contains enough material for a short introductory course for students at an undergraduate level. You will find some sections of this book marked with a star ★ at the beginning of the title. That means that the section is more mathematically challenging and can be safely skipped by those with less mathematical know-how or interest.

This book is structured to provide an accessible overview of statistics and data analysis for scientific research. It begins with basic concepts, including an explanation of random variables, outcome spaces, and the difference between descriptive and inferential statistics. It continues with data types, measures of central tendency, of dispersion, and of positions. The discussion continues with a discussion of outliers and various methods to define them. Then the book introduces more complex topics like distributions, hypothesis testing, and regression analysis. Each chapter builds on the previous one, introducing more complex statistical techniques in a step-by-step manner, making it suitable for readers ranging from beginners to those needing a quick refresher. The choice of topics, as often, is somewhat subjective. But I selected the concepts that all those interested in using statistics should know about. Some topics (such as hypothesis testing) are not discussed at length, since a complete treatment of the topics would require not only more space but also the student having some more background in statistics and mathematics (e.g., in calculus). The goal of these chapters is to give the student enough understanding of the main idea and inspire them to study the concepts deeper.

In the book you will find four types of boxes: *definitions*, *tips*, *warnings*, and *examples*. The meaning should be clear. In *definitions* boxes, you will find definitions of concepts, so it is easier to find them. In *tips* you will find, as the name suggests, tips and suggestions that I hope will make specific concepts or applications clearer. In *warnings* I try to highlight tricky cases in the application or interpretation of methods or concepts. And finally, in *examples* I try to give some examples to make concepts clearer.

Two books have significantly influenced my journey in learning statistics. One is the book *Probability and Statistical Inference* by Hogg, Tanis, and Zimmerman now in its 10th edition [1]. In my opinion, it is the perfect book for (almost) *beginners*. A more advanced book, but beautiful in its rigor and choice of topics, is the one by Casella and Berger *Statistical Inference* [2]. A masterpiece that every scientist should have on his or her bookshelf. If you are looking to delve deeper into statistics after reading this short introduction, you cannot go wrong with those two books. If you are a beginner, after this book, I suggest you continue with the book by Hogg et al.

Dübendorf, Switzerland February 2025	Umberto Michelucci

Declarations

Competing Interests The author has no competing interests to declare that are relevant to the content of this manuscript.

Contents

1 Introduction to Statistics 1
   1.1 Household Budget Survey 1
   1.2 A Brief Introduction to Statistics 2
   1.3 Random Experiments, Random Variables, Outcome Space and Events 8
   1.4 Descriptive and Inferential Statistics 11
   1.5 Descriptive Statistics 12
   1.6 Inferential Statistics 13
   1.7 Data Analysis vs. Statistics 14

2 Types of Data 15
   2.1 Qualitative (Categorical) Data 15
   2.2 Quantitative (Numerical) Data 16
   2.3 Level of Measurements 17
   2.4 Cohort 18
   2.5 Longitudinal Data 19
   2.6 Cross-Sectional Data 20
   2.7 Binary and Dichotomous Data 20

3 Data Collection Methods (Sampling Theory) 23
   3.1 Introduction 23
   3.2 Research Questions and Hypotheses 24
      3.2.1 Research Questions 24
      3.2.2 Hypothesis 25
   3.3 Survey Sampling 26
      3.3.1 Non-probability Sampling 28
      3.3.2 Probability Sampling 29
   3.4 Stratification and Clustering 30
   3.5 Random Sampling Without Replacement 31
   3.6 Random Sampling with Replacement 32

		3.7	Random Stratified Sampling	33
		3.8	Bootstrap	34
4	**Measures of Central Tendency**			**39**
	4.1	Introduction		39
	4.2	Mean		39
	4.3	Median		42
	4.4	Mode		43
	4.5	Mid-Range		44
	4.6	When to Use Mean, Median or Mode		46
5	**Measures of Dispersion**			**47**
	5.1	Variance		47
	5.2	Standard Deviation		49
	5.3	Range		51
	5.4	Dangers of Relying on Single Statistics		52
6	**Measures of Position**			**55**
	6.1	Introduction		55
	6.2	Percentiles		56
		6.2.1	Nearest-Rank Method	57
		6.2.2	Linear Interpolation Between Ranks	57
	6.3	Quartiles		62
	6.4	Interquartile Range		63
	6.5	Deciles		64
	6.6	Quantiles		65
7	**Outliers**			**67**
	7.1	Introduction		67
	7.2	Interquartile Range (IQR) Method		67
	7.3	Domain-Specific Criteria		68
	7.4	z-Score Method		68
	7.5	Causes, Impact, and Treatment		70
8	**Introduction to Distributions**			**73**
	8.1	A Small Warning		73
	8.2	Introduction to Probability Distributions		73
		8.2.1	Discrete Probability Distribution	74
		8.2.2	Continuous Probability Distribution	75
		8.2.3	Cumulative Distribution Function	77
		8.2.4	Expected Value and Variance	77
	8.3	The Normal Distribution		78
	8.4	★ Mathematical Description of the Normal Distribution		80
		8.4.1	★ Significance of μ and σ	81
	8.5	Bernoulli Distribution		86
	8.6	Binomial Distribution		88
	8.7	The Poisson Distribution		89
	8.8	Probability Distributions: An Overview		90

9 Skewness, Kurtosis, and Modality ... 95
- 9.1 Characteristics of a Distribution ... 95
- 9.2 Skewness ... 96
  - 9.2.1 Pearson's Skewness Coefficients ... 99
  - 9.2.2 Quantile-Based Skewness Measures ... 101
  - 9.2.3 Further Ways of Measuring Skewness ... 101
- 9.3 ★ Kurtosis ... 102
- 9.4 Modality ... 103
- 9.5 ★ Moments of a Distribution ... 104
- 9.6 ★ Central Moments ... 107

10 Data Visualisation ... 109
- 10.1 Histograms ... 109
- 10.2 Boxplots ... 112
- 10.3 *Q-Q* Plots ... 115
- 10.4 Pair Plots ... 117

11 Confidence Intervals ... 121
- 11.1 Introduction ... 121
- 11.2 Confidence Intervals for the Mean ... 122
- 11.3 Bootstrap Confidence Intervals ... 126

12 Hypothesis Testing ... 129
- 12.1 Disclaimer ... 129
- 12.2 Hypothesis Testing: The Basic Idea ... 129
- 12.3 An Example ... 130
- 12.4 Test of One Mean: Variance Known ... 132
- 12.5 Test of One Mean: Variance Unknown ... 132
- 12.6 *p*-Values: An Intuitive Definition ... 134
- 12.7 Type I and Type II Errors in Hypothesis Testing ... 135

13 Correlation and Linear Regression ... 137
- 13.1 Correlation ... 137
- 13.2 Regression Analysis ... 140
  - 13.2.1 Linear Regression ... 141
  - 13.2.2 Coefficient of Determination ... 142
- 13.3 Further Readings ... 144

14 Ethics and Best Practices ... 145
- 14.1 Steps of a Statistical Project ... 145
- 14.2 Reproducibility, Replicability, Transparent Reporting, and Documentation ... 147
  - 14.2.1 Reproducibility ... 148
  - 14.2.2 Data and Code Sharing ... 150
  - 14.2.3 Transparent Reporting ... 151
  - 14.2.4 Best Practices for Documentation ... 152

Glossary 155

A ★ **Partitioning of the Ordinary Least Square Variance** 157

B **Big-O and Little-o Notation** 161
    B.1 Big-O Notation 161
    B.2 Little-o Notation 162

References 163

Index 165

List of Definitions

1.2.1: Population 2
1.2.2: Sample 2
1.3.1: Random Experiments, Outcome Space, and Events 8
1.3.2: Random Variable 10
1.4.1: Realisation of a Random Variable X 11
2.1.1: Qualitative (Categorical) Data 15
2.2.1: Quantitative (Numerical) Data 17
2.3.1: Nominal Level of Measurement 17
2.3.2: Ordinal Level of Measurement 17
2.3.3: Interval Level of Measurement 18
2.3.4: Ratio Level of Measurement 18
2.4.1: Cohort 18
2.5.1: Longitudinal Data 19
2.6.1: Cross-Sectional Data 20
2.7.1: Binary Data 20
2.7.2: Dichotomous Data 20
3.2.1: Research Question 25
3.2.2: Hypothesis 26
3.3.1: Survey Sampling 27
3.3.2: Non-probability Sampling 28
3.3.3: Probability Sampling 30
3.4.1: Stratified Population 30
3.4.2: Clustered Population 31
3.5.1: Random Sampling Without Replacement 32
3.6.1: Random Sampling with Replacement 33
4.2.1: Mean (Average) 40
4.3.1: Median 42
4.4.1: Mode 44
4.5.1: Mid-Range 45
5.1.1: Variance 47
5.2.1: Standard Deviation 50

5.3.1: Range 51
6.2.1: Percentiles 62
6.3.1: Quartiles 62
6.4.1: Interquartile Range 64
6.5.1: Deciles 65
7.2.1: Outliers Defined with the IQR 68
7.4.1: Outliers Defined with the z-Score 69
8.2.1: Probability Distribution (Intuitive Definition) 73
8.2.2: Probability Mass Function 74
8.2.3: Probability Density Function 75
8.2.4: Cumulative Distribution Function 77
8.3.1: Normal Distribution 80
8.7.1: (Approximate) Poisson Process 89
9.1.1: Tails of a Distribution 95
9.1.2: ★ Heavy Tail Distribution 96
9.2.1: Skewness 97
9.2.2: Pearson Mode and Median Skewness Coefficients 100
9.5.1: Moment Generating Function (mgf) 107
12.7.1: Type I Error 135
12.7.2: Type II Error 135
13.1.1: Pearson Coefficient r 137
13.1.2: Covariance 138
13.2.1: Trend 141

List of Tips

1.2.1: You Want to Publish on Nature? 7
1.3.1: ★ Countable Random Variable 9
3.3.1: Meaning of Survey 27
4.2.1: Expected Value 40
4.6.1: When to Use Mean, Median, or Mode 46
5.1.1: ★ N or $N-1$ in the Variance Formula? 48
5.2.1: Sample Variance S^2 51
6.2.1: Virtual Index Notation 58
8.4.1: ★ Expected Value of a Symmetric Distribution 82
8.4.2: ★ Proof of Eq. (8.10) 82
8.4.3: ★ Proof that $A = 1$ in Eq. (8.23) 84
9.1.1: Heavy and Light Tails (Intuitive Explanation) 96
9.2.1: ★ $\gamma_1 = 0$ for a Symmetric Distribution 97
9.3.1: ★ Proof Sketch that $\mathbb{E}[(X-\mu)^4] = 3\sigma^4$ for a Normal Distribution 102
9.5.1: Moments of a Distribution 105
9.5.2: ★ Proof that $\mathbb{E}(X^n) = M_X^{(n)}(0)$ 106
10.1.1: Practical Tips for Building a Histogram 111
10.1.2: Determination of the Number of Bins in Histograms 112
10.2.1: Outliers in Boxplots 114
10.2.2: When to Use Boxplots 114
10.3.1: Creating Q-Q Plots in Python 115
10.3.2: Finding the Best Fitting Distribution in Python 117
10.4.1: The Iris Dataset 117
10.4.2: How to Read a Pair Plot and Why to Use It 119
11.2.1: Meaning of Confidence Interval 123
11.2.2: Central Limit Theorem 124
11.2.3: Confidence Intervals for Difference of Mean, Proportions and More ... 125
11.3.1: Bootstrap in a Nutshell 126

12.3.1: Calculation of p-Value 131
12.5.1: Hypothesis Testing in Practice 133
13.2.1: Regression Beyond Linear Cases 140
13.2.2: Regression 141
13.2.3: R^2 Use Tips 143

List of Warnings

1.2.1: Why Statistics Is Sometimes Poorly Regarded 5
3.5.1: Random Sampling Without Replacement 31
3.6.1: Random Sampling with Replacement 32
5.4.1: Dangers of Relying on Single Statistics 52
6.2.1: Percentiles in numpy 59
6.3.1: Calculations of Quartiles 63
7.4.1: z-Score for Not Normally Distributed Data 70
8.2.1: PDF and the Probability of a Specific Value 75
9.2.1: Relationship Between Mean and Median 99
9.2.2: Pearson Skewness Coefficients 101
10.2.1: Limitations of Boxplots 114
12.5.1: Hypothesis Testing: More Complex Cases 133
13.1.1: The Pearson Coefficient and Non-linear Relationships.............. 139
14.1.1: Cofounding Variables.. 146
14.2.1: Reproducibility and Replicability 149

List of Examples

1.2.1: Main Goal of Statistics 3
1.2.2: Bad Statistics 3
1.3.1: Random Variable 10
1.5.1: Descriptive Statistics 12
1.6.1: Inferential Statistics 13
1.7.1: Data Analysis 14
2.1.1: Qualitative (Categorical) Data 15
2.2.1: Quantitative (Numerical) Data 16
2.4.1: Cohort 19
2.5.1: Longitudinal Data 19
2.6.1: Cross-Sectional Data 20
2.7.1: Non-dichotomous vs. Binary Data 21
4.2.1: Mean of an Array of Numbers 41
4.3.1: Median of an Array of Numbers 43
4.4.1: Mode of an Array of Numbers 44
4.5.1: Mid-Range of an Array of Numbers 45
5.2.1: Variance and Standard Deviation 50
6.2.1: Percentiles with the Nearest-Rank Method 57
6.2.2: Percentiles 60
6.3.1: Quartiles 62
7.3.1: Domain-Specific Criteria for Outliers 68
7.4.1: z-Score 69
8.2.1: Probability Mass Function: Rolling a Six-Face Dice 75
8.2.2: PDF and Why Integration Is Needed 76
8.3.1: Normal Distribution 79
8.5.1: Bernoulli Trials 86
8.5.2: Bernoulli Distribution 87
8.6.1: Bernoulli and Binomial Distribution 88
8.7.1: Poisson Distribution 90
9.2.1: Right (Positively) Skewed Distributions 98
9.2.2: Left (Negatively) Skewed Distributions 98

9.4.1: A Bimodal Distribution 104
10.1.1: Histogram of Test Scores 110
11.1.1: Sample and Population Averages 121
11.2.1: Confidence Interval 123
11.3.1: Bootstrap for Confidence Interval Evaluation 126

Acronyms and Symbols

ANOVA   Analysis of Variance: a collection of statistical models to compare multiple group means.

AUC   Area Under the Curve: used in classification analysis to determine which of the used models predicts the classes best.

CI   Confidence Interval: a range of values used to estimate the true value of a population parameter.

IQR   Interquartile Range: a measure of statistical dispersion, being equal to the difference between 75th and 25th percentiles, or between upper and lower quartiles.

LR   Logistic Regression: a statistical model that in its basic form uses a logistic function to model a binary-dependent variable.

MLE   Maximum Likelihood Estimation: a method of estimating the parameters of a statistical model.

N   Sample Size: the number of observations in a sample.

p-value   Probability Value: the probability of obtaining test results at least as extreme as the ones observed during the test, assuming that the null hypothesis is correct.

R   R (Programming Language): a language and environment for statistical computing and graphics.

ROC   Receiver Operating Characteristic: a graphical plot that illustrates the diagnostic ability of a binary classifier system.

RSS   Residual Sum of Squares: a measure of the discrepancy between the data and an estimation model.

SAS   Statistical Analysis System: a software suite used for advanced analytics, multivariate analyses, business intelligence, and data management.

SD, σ Standard Deviation: a measure that is used to quantify the amount of variation or dispersion of a set of data values.

SEM, $\sigma_{\bar{X}}$ Standard Error of the Mean: an estimate of how far the sample mean is likely to be from the population mean.

SPSS	Statistical Package for the Social Sciences: a software package used for statistical analysis.
TSS	Total Sum of Squares: sum of squares between the data and an estimation model.

Chapter 1
Introduction to Statistics

> *To call in the statistician after the experiment is done may be no more than asking him to perform a post-mortem examination: he may be able to say what the experiment died of.*
> – *Sir R.A. Fisher (1938)*

1.1 Household Budget Survey

To understand the main goal of statistics, consider this example: Since 1990, the Swiss federal statistical office has tried to estimate the typical household budget (HB) in Switzerland every year. Although the most accurate method would be to ask every household about their budget (income and expenses), this approach is clearly impractical. The sheer number of households makes it difficult to contact all of them, and many would likely be unwilling to share their financial information. Therefore, a different approach is necessary. The statistics office conducts the Household Budget Survey (HBS), in which approximately 3000 households across Switzerland are surveyed. Statistical methods come into play by allowing the office to extrapolate the information from these 3000 households to make estimates about *all* households in Switzerland. Not only that, statistics also helps in the survey design, by providing methods to keep into account differences between, for example, rural and cities (it is to be expected that people in cities earn more than people in rural areas) or between different job types (e.g. bankers and farmers). Furthermore, statistics helps in choosing the most appropriate sampling method, to ensure that the sample is representative of the entire population. After data collection, statistics allows for correction for any over-represented or under-represented groups in the sample. For example, if larger households are less likely to respond to the survey (or are generally less than small households), statistical methods can adjust the results to better reflect the true distribution of household sizes in the population. If some households do not respond to certain questions or the entire survey, statistical techniques such as imputation[1] can be used to estimate the missing data based on the available information. Statistics can be used to test hypotheses, such as whether there is a significant difference in household budgets between different regions or income groups. This helps confirm whether the observed differences in the sample are likely to be true for the entire population.

---

[1] Imputation is a statistical technique used to estimate and replace missing data within a dataset.

In conclusion, statistics provides the tools necessary to make reliable estimates about a population based on a sample (more on those terms later), ensuring that the data collected are representative and adjusted for any biases. It allows analysis of patterns, testing of hypotheses, and correction of missing or incomplete information. Through these methods, statistics enables informed decisions and accurate insights into larger populations from limited data.

1.2 A Brief Introduction to Statistics

The primary goal of statistics is to infer characteristics of a larger population based on experimental observations from a subset (called a **sample**) of that population. Here, the term **population** refers to the complete set of data points relevant to a specific problem (more on that in Chap. 3), while a **sample** is a representative segment of that population.

> *Definition* 1.2.1: **Population**
>
> In statistics, a **population** refers to the entire set of individuals or objects of interest that share at least one common characteristic (typically multiple). It is the complete group that researchers are interested in studying and from which they often draw samples for analysis. The population can be finite (e.g. all students in a particular school) or infinite (e.g. all possible outcomes of rolling a die infinite times).
> Two examples of population are all patients with type 2 diabetes in a particular country or all doctors in a city.

Access to an entire population is typically not possible. For instance, it is unrealistic to expect access to every individual worldwide who has a specific heart disease. Therefore, we rely on a sample, hoping to get enough information about the entire population from it.

> *Definition* 1.2.2: **Sample**
>
> In statistics, a **sample** is a subset of individuals or objects selected from a larger population. The sample is used to make inferences or generalisations about the population from which it was drawn. For example, to study the dietary habits of adults in a city, a sample of 1,000 adults from the city's population may be used, since attempting to survey every adult in the city is impossible.

1.2 A Brief Introduction to Statistics

Example 1.2.1: **Main Goal of Statistics**

Imagine you are working in a school with 1000 students (the **population**). You want to assess the quality of teaching and devise a test for students. Since you cannot ask 1000 students to perform the test, you choose 50 of them randomly (the **sample**). Your assumption is that the results that you obtain from 50 will be representative of the 1000. Statistics helps you by giving you the means to do that (for example, it studies how to choose the 50 students from the 1000 to avoid hidden bias, how to describe data properly, and much more) and to assess your results in relation to the entire population (the 1000 students).

Statistics is also crucial for comparing groups and determining whether there are significant[2] differences between them. For example, questions like whether women are generally shorter than men or whether students from one school perform better than those from another can be answered using statistical methods. But it is important to frame these questions precisely since, while general trends may exist, there will always be exceptions, such as some women being taller than some men. Accurate hypothesis formulation, precise language, and proper result communication and visualisation are fundamental to statistical analysis, as we will discuss at length in this book.

Example 1.2.2: **Bad Statistics**

Christopher Engledowl and Travis Weiland discussed misleading data visualisations in their article "Data (Mis)representation and COVID-19: Leveraging Misleading Data Visualisations For Developing Statistical Literacy Across Grades 6–16" [3]. They highlighted a specific case involving a chart by the US Georgia Department of Public Health. This chart, published in May 2020, intended to show the top five counties with the highest COVID-19 cases over the past 15 days, as well as the progression of the cases over time. The chart they published is reproduced in the top panel in Fig. 1.1. Upon examining the chart, where a clear downward trend is visible, several significant issues were noted that contributed to its misleading nature.
Firstly, the x-axis lacked labels, which was critical since it was supposed to track the number of cases over time. Moreover, the sequence of dates under the bars was **not** chronological. The dates of April and May dates were mixed, seemingly suggesting a decrease in cases. Furthermore, the order of the counties varied, each being presented in descending order of cases to further imply that the situation was improving.

---

[2] Significant in this context means due to other reasons other than luck.

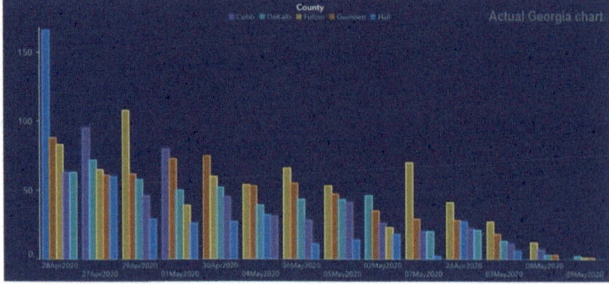
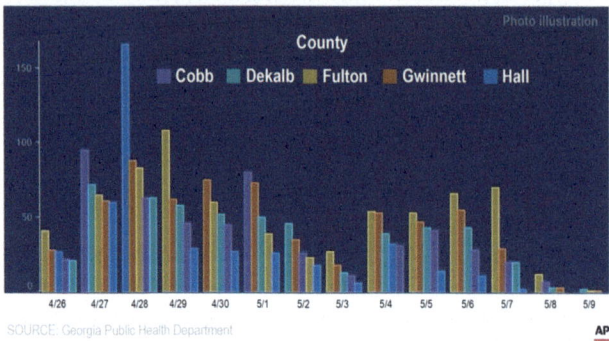

Fig. 1.1: Figure reproduced from [4]. The plot at the top shows the highly misleading plot published by the Georgia Department of Public Health and the one at the bottom shows the corrected one. Quite a difference, clearly

This graph stirred considerable controversy, particularly on social media platforms such as Twitter, where users criticised the Georgia Health Department for using misleading statistics during the pandemic. In response, Candice Broce, the communications director for Georgia's Governor Brian Kemp, admitted that the arrangement of the x-axis was intended to highlight descending values to more clearly show peak values and the affected counties on those dates. She acknowledged that the graph did not meet its intended purpose and apologised for the confusion caused. The graph was subsequently corrected to display the dates and counties in their proper order. To

1.2 A Brief Introduction to Statistics

see the corrected version, check the bottom diagram in Fig. 1.1 reproduced from [4]. The difference with the misleading one is quite striking.
Other examples of bad statistics are related to bias in data sampling, tendentious communication, and many more. Note that statistics is **not** guilty, scientists and bad science and practices are.

I do not agree with what Churchill said, namely "I only believe in statistics that I doctored myself". If a study is well documented, the hypotheses well stated, and all the information available, one can and should trust the statistical results of the study itself. But, alas, sometime statistics has a bad name. Its negative reputation often comes from a lack of understanding about how it operates (from researchers and the public), coupled with the tendency to publish results without providing all necessary information to assess the validity of those results. If you read about a study, you should always check if the researchers gave all the necessary information about the analysis (hypothesis, data collection strategy, experimental design, etc.). If this information is not available, it is impossible to judge the validity of the study, and its conclusions should be challenged or even ignored if necessary.

Warning 1.2.1: **Why Statistics Is Sometimes Poorly Regarded**

There are several reasons for which statistics sometimes gets a bad name. But probably the most important ones are the following.
Misleading presentation: This is a big one. Statistics can be presented in ways that lead audiences to incorrect conclusions. This includes using graphs with distorted scales (to highlight or underline relative sizes, for example), displaying only certain portions of the data (you only show the data that support your message) or presenting data without context. When statistics are visualised misleadingly, they can exaggerate trends or mask important details.
Cherry picking data: This involves selectively presenting data that support a specific conclusion while ignoring data that contradict it (you can ignore outliers or, with mischievous intentions, ignore good data). Cherry picking can give a skewed view of reality and is often used to persuade or mislead rather than inform. When presenting your study and your data you **must** always present all your data, including outliers or results that may seem strange or not in line with your hypothesis.
Correlation mistaken for causation: A common mistake in the interpretation of statistics is the assumption that the correlation between two variables implies that one causes the other. This mistake can lead to wrong beliefs about relationships between factors and outcomes, often oversimplifying complex interactions (or simply getting them wrong). If you want to have fun, check the website by Tyler Vigen at [5] (https://www.tylervigen.com/spurious-correlations), where he collects spu-

rious correlations. For example, check Fig. 1.2. I think we all agree that there is no causality or correlation between the Kerosene used in South Korea and Google searches for "report UFO sighting". This is a very good example of a spurious correlation.

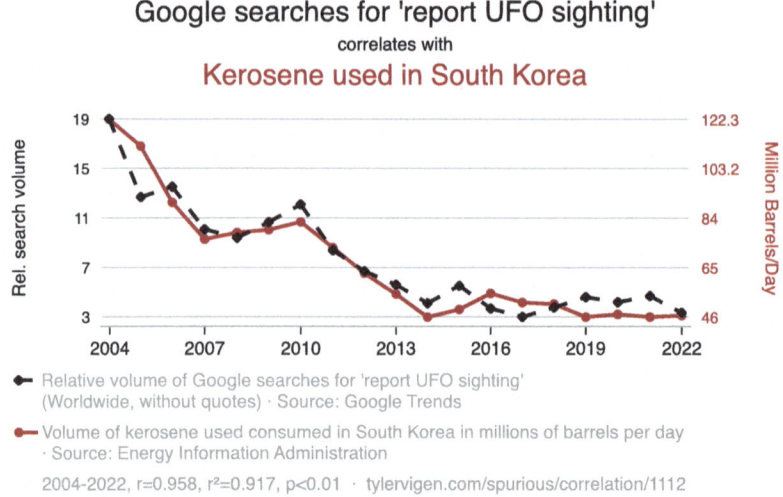

Fig. 1.2: Figure reproduced from [5]. I think we all agree that there is no causality or correlation between the kerosene used in South Korea and Google searches for "report UFO sighting". This is a very good example of spurious correlation

You must be very careful when assuming that correlation (more on this in Sect. 13.1) is the same as causation.

Lack of reproducibility: Bad statistics may arise from studies or analyses that cannot be replicated by others. This may be due to inadequate methodology or missing documentation on the study itself (you will find numerous papers with studies that are not documented and thus not reproducible). Statistics that cannot be reproduced are **not** credible and can contribute to a lack of trust in their statistical conclusions.

Biased data collection: When data are collected with bias or the sample is not representative of the population, the resulting statistics will inherently reflect that bias. This can mislead decision-making processes and policy formulations, leading to ineffective or harmful outcomes based on incorrect data interpretations. Consider the example of exit polling. In this method, volunteers approach individuals leaving a polling station to enquire about their voting choices. However, this approach automatically excludes absentee voters and may suffer from bias in data collection. Research indicates that volunteers, often younger, college educated, and predominantly white, may subconsciously choose to interact with people who resemble themselves, such as fellow students, over potentially different demographics such as a middle-aged parent managing several children. As a result, not everyone has the same likelihood of being selected for an exit poll, leading to potential biases in the data gathered.

Tip 1.2.1: **You Want to Publish on Nature?**

We all know that Nature is one of the most important scientific journals. If you think that concepts such as those in this book are not truly necessary, you should think twice. When you submit a paper to Nature, you must fill out a form[a] with questions (some examples are reported below) asking you to confirm if you provided the following in your paper.

- *A description of any assumptions or corrections, such as tests of normality and adjustment for multiple comparisons*
- *A full description of the statistical parameters including central tendency (e.g. means) or other basic estimates (e.g. regression coefficient) AND variation (e.g. standard deviation) or associated estimates of uncertainty (e.g. confidence intervals)*
- *For null hypothesis testing, the test statistic (e.g. F, t, r) with confidence intervals, effect sizes, degrees of freedom, and P value noted*
- *The statistical test(s) used AND whether they are one- or two-sided. Only common tests should be described solely.*

If the questions are unclear to you, it is likely that your study is not well designed or executed. Note that the questions are not strange or an exaggeration, but they are the minimum of information that any statistical study should give. At this point in the book, you probably do not understand the questions, but at the end of the book you should be able to see why they are important and what they mean.

---

[a] The form is licensed under a Creative Commons Attribution 4.0 International License, which permits use, sharing, adaptation, distribution, and reproduction in any medium or format, as long as you give appropriate credit to the original author(s).

1.3 Random Experiments, Random Variables, Outcome Space and Events

As Wittgenstein said in his Tractatus logigo-philosphicus, *the limit of my language means the limit of my world*, so we need to clarify some terminology before proceeding to fence off misunderstandings. Generally speaking, in statistics we deal with experiments for which the outcome cannot be predicted with certainty, such as the tossing of a coin or a dice or measuring the height of a person randomly chosen. These are called **random experiments**. The term random indicates that the outcome cannot be predicted with certainty. Furthermore, we speak of an **outcome space** S, that is, the collection of all possible outcomes of a random experiment. For example, for the tossing of a six-face dice, the space S is the set $S = \{1, 2, 3, 4, 5, 6\}$. Of course, some measurements can be of continuous variables, like the height of a person. If we consider that adult human heights generally range from about 120 to 250 cm, the outcome space S for a random experiment measuring the height of adults can be defined as the interval [120 cm, 250 cm].

An **event** is a subset of the outcome space S. An event can be, in the example of the dice, the number 5 (you toss the dice and get a 5). The number 5 is a subset of $S = \{1, 2, 3, 4, 5, 6\}$. In more intuitive terms, an event is nothing else than something that can arise from a random experiment.

Definition 1.3.1: **Random Experiments, Outcome Space, and Events**

Here are the definitions of the concepts we just discussed.

Definition 1.1 (Random Experiment) Experiments for which the outcome cannot be predicted with certainty are called **random experiments.**

Examples are tossing a coin or a dice or measuring the height of a person randomly chosen.

Definition 1.2 (Outcome Space S) The collection of all possible outcomes of a random experiment is called the outcome space and is typically indicated by S.

Imagine that you have a standard deck of 52 playing cards, which includes 13 cards each of four suits (clubs ♣, diamonds ◊, hearts ♡, and spades ♠). Each suit contains cards numbered from 2 to 10, plus a Jack, Queen, King, and Ace. When you draw a single card from the deck, the outcome space S is the set of all possible cards you could draw. Thus, the outcome space *for the random experiments of picking up one single card* includes 52 elements, each representing a unique card.[a]

---

[a] Remember that an outcome space is only defined for a given random experiment.

1.3 Random Experiments, Random Variables, Outcome Space and Events

Definition 1.3 (Event A) An event A is a subset[b] of the outcome space S.

To be a bit clearer, let us consider an example. Suppose you have a six-face fair dice. The random experiment we consider will be the throwing of the dice together with the observation of the number that is on the top face when the dice come to rest on the surface you have thrown the dice on. Now, since the dice will not disappear mid-air, one of the six numbers must appear, so the outcome space will be $S = \{1, 2, 3, 4, 5, 6\}$. For example, the event of getting a 6 would be $A = \{6\}$, or the event of getting 2 will be $A = \{2\}$. Note that all the events described are subsets of S. An event cannot contain something that is not in the outcome space. In our example, $\{7\}$ is not an event as will never happen since the dice has only six faces. An event can also contain multiple elements, depending on the random experiment. Consider, for example, the experiment of throwing two dice at the same time. The outcome space will now be larger

$$S = \{(1,1), (1,2), \ldots, (1,6), (2,1), \ldots, (6,6)\} \tag{1.1}$$

and will contain 36 elements. The event of getting the same number on both dices will be a set of six elements.

$$S = \{(1,1), (2,2), (3,3), (4,4), (5,5), (6,6)\} \tag{1.2}$$

It is important to note that the events and outcome spaces depend on the random experiment you are performing.

---

[b] A subset is a portion of a given set of elements. For example, if you have the set of integers from 1 to 10 $\{1, 2, 3, 4, 5, 6, 7, 8, 9, 10\}$, examples of subsets of it could be the even $\{2, 4, 6, 8, 10\}$ or the odd integers $\{1, 3, 5, 7, 9\}$.

One fundamental concept in statistics is that of a **random variable**, that is, a function that associates with each event a number. There are two types of random variables: **discrete** and **continuous**. A **discrete random variable** is one that can assume a countable number of distinct values. "Countable" here means that the values can be listed, like rolling a die (with outcomes 1, 2, 3, 4, 5, and 6) or counting the number of cars passing through an intersection in a day.

Tip 1.3.1: ★ Countable Random Variable

A random variable X is said to be **countable** if its range, denoted $\mathcal{R}(X)$, is a countable set. $\mathcal{R}(X)$ is countable, if there exists a bijective function $f : \mathcal{R}(X) \to \mathbb{N}$, where $\mathbb{N}$ denotes the set of natural numbers. Intuitively, this means that to each of the elements of the range of X you can uniquely assign a natural number.

The key characteristic of a discrete random variable is that there are gaps between the possible values it can take; it does not cover a continuous span of values (like the height of a person, for example). A **continuous random variable**, on the other hand, can take on an infinite number of possible values within a given range. These values are uncountable and can include every possible value between two numbers. For example, the height of a person can be considered as a continuous random variable because it can take any value within a range, such as, for example, from 150 to 190 cm, and this can include measurements like 176.52467 cm, which can be as precise as you wish them to be (at least within measurement errors of course).

> *Definition* 1.3.2: **Random Variable**
>
> A **random variable** is a function that associates a number with each event in a random experiment. There are two types of random variables: **discrete** and **continuous**.

In our example of tossing two dice, a random variable could be the sum of the numbers coming out for each roll. Or, if you consider tossing just one dice, a random variable could be the sum of the results of 50 tosses. Usually, a random variable is indicated with an uppercase letter such as X or Y. A random variable is called "variable" because it represents a value that can vary due to chance. The term "variable" emphasises that, unlike a constant, the value it takes is not fixed but can change depending on the outcome of the random process with which it is associated.

This interpretation of the name is something not everyone agrees on. The sentence *a random variable is neither random nor a variable* is attributed (but a specific reference has eluded me) to Giancarlo Rota[3] and highlights how the interpretation of the name is not appreciated by everyone. The important thing to remember is that a random variable is a function that associates a number with an event.

> *Example* 1.3.1: **Random Variable**
>
> **Discrete Random Variable**: Let X be a discrete random variable that represents the number of heads when flipping three coins. The possible values of X are
>
> $$X \in \{0, 1, 2, 3\}$$
>
> **Continuous Random Variables**: Let Y be a continuous random variable representing the time it takes, in hours, for a chemical reaction to complete. Y can take any value in the interval:
>
> $$Y \in [0, \infty)$$
>
> Of course, ∞ is an exaggeration, and probably a value of several hundreds of hours would suffice.

---

[3] This attribution can be found in [6].

1.4 Descriptive and Inferential Statistics

If your project involves data,[4] you should always start by understanding it. The distribution (more in Chap. 8) of your data (data distribution refers to the way in which values in a dataset are spread or dispersed, showing how often each value or range of values occurs) may be quite complex, but generally you want to have a rough idea about it. Understanding your data is what is addressed by **descriptive statistics**. This definition is somewhat intuitive and is better discussed in Sect. 1.5.

> *Definition* 1.4.1: **Realisation of a Random Variable** X
>
> The **realisation** of a random variable X is the value that is actually observed in an experiment. The realisation of X is often called **observation** or **observed value**.

Suppose that you have a random variable X, and measure it N times (we will indicate with x_i, $i = 1, \ldots, N$ the N realisations (see Definition 1.4.1) of X). The questions you should try to answer are, at least, the following:

- What is the *typical* value of X? For example, if you assess the grades of students in a school, the first thing that is important to know is what the *typical* grade of students in all disciplines is. What *typical* means is somewhat subjective, and we will discuss it later when we discuss **measures of central tendency** (see Chap. 4), which are used to answer exactly this question.
- How are the data spread around its typical value? If we continue to consider the examples of students, it is interesting to know if the grades go, say, on a scale from 0 to 100, from 10 to 100 or from 70 to 90. In fact, the former case may indicate problems with students or teaching; the latter does not point to problems (at least not to problems easy to detect). For this question, you use **measures of dispersions** (see Chap. 5).
- How are the data spread over its range? Suppose that you really find that your student grades go from 10 to 100. Is this really a problem? If only one student has 10 and all other students have grades that start at 70, then the school has no problem. If 50% of the student grades are less than 15, the school probably has problems of some form. So how the data are spread over their range (in this example, the range [0,100]) is quite important, and let us better understand the data. To study this, you use **measures of position** (see Chap. 6).

After understanding your data, you may want to predict something about your population from your data (the sample). In our example of students, maybe you do not have the grades of all the students, but only of 10% of them, since it was impractical to get the grades for everyone. To study the sample (the 10% of student grades) and infer properties of the population, techniques that go under the name of **inferential statistics**, discussed in detail in Sect. 1.6, should be used. Here are some examples

---

[4] And if you are doing statistics, your project **will** always involve data.

of questions you may be interested in, which fall under the umbrella of inferential statistics.

- To determine if a new teaching method is effective, you compare the average grades of students taught with the new method to those taught with the traditional method. Hypothesis testing (see Chap. 12) may support the hypothesis that students who use the new method have significantly higher grades, suggesting that the method is effective.
- To estimate the average grade of all students in the school, you calculate a confidence interval (see Chap. 11) based on the grades of a random sample of students. This interval provides a range in which the true average grade is likely to fall, giving an estimate of overall academic performance with **confidence** of this estimate.
- To understand the factors that influence student grades, you use regression analysis (see Sect. 13.2) to examine the relationship between grades and, for example, variables such as study hours, attendance, and parental involvement. The analysis may reveal, for example, that study hours and attendance are strong predictors of higher grades.

1.5 Descriptive Statistics

Descriptive Statistics goals are summarising and organising data so that it can be understood and presented in a meaningful way. It typically provides summaries of the characteristics about a given data sample.

Typically descriptive statistics is divided into three groups.

- **Measures of central tendency (mean, median, and mode)** which describe the *centre* of the data (see Chap. 4), or in other words the "typical" characteristics of data
- **Measures of variability (variance, standard deviation, and range)** which describe the variability or dispersion within the data (see Chap. 5)
- **Measures of position (percentiles and quartiles)** which provide insights into the distribution of data across different intervals (see Chap. 6)

> *Example* 1.5.1: **Descriptive Statistics**
>
> Here are some examples of cases where descriptive statistics is used. Note that the concepts outlined intuitively here will be explained in detail in Chaps. 4, 5, and 6.
>
> **Customer satisfaction survey analysis**: A company sends out a survey to its customers to rate their satisfaction on a scale of 1 to 5, with 5 being the highest level of satisfaction. The company collects the responses and calculates the average (mean) score, the most frequently selected score (mode), and the spread of scores (standard deviation) to understand overall customer satisfaction and areas of improvement.

Annual income report for a region: The government collects data on the annual incomes of individuals within a specific region and reports the average income (mean), the income level that divides the population into two equal halves (median), and the income range (from the lowest to the highest). This information helps to understand the economic status of the region.

High school test scores: A high school gives a standardised test to its students. After grading the tests, the school calculates and reports the average score (mean), the score in the middle of the dataset (median), and the difference between the highest and lowest scores (range). These data help to assess the overall performance of the students and identify areas where students may need additional help or resources.

1.6 Inferential Statistics

Inferential Statistics objective is to make inferences or predictions about a population from which a sample was drawn. It is used to make judgements about the probability that, for example, an observed difference between groups is real or that it might have happened by chance. Thus, inferential statistics allow us to infer the properties of a population based on a sample. This includes hypothesis testing, confidence intervals, and regression analysis.

Example 1.6.1: **Inferential Statistics**

Here are some examples of cases where inferential statistics is used. The concepts described intuitively here can be found in Chaps. 11, 12, and 13.

Predicting Election Outcomes: A political analyst uses data from pre-election polls to predict the outcome of an upcoming election. By applying inferential statistics, the analyst can estimate the proportion of the population that supports each candidate, along with a confidence interval for these proportions, to make an educated guess about the winner.

Drug Efficacy in Clinical Trials: In a clinical trial, a pharmaceutical company wants to determine whether a new drug is more effective than existing treatments. Using inferential statistics, researchers compare the health outcomes of patients who use the new drug with those using a placebo or the current standard treatment. Statistical tests help determine whether any observed differences are statistically significant.

Market Research for Product Launch: Before launching a new product, a company conducts market research to understand the preferences of potential customers and willingness to buy. By collecting data from a sample of the target market and applying inferential statistics, the company can make generalisations about the entire target market's preferences and likely purchase behaviour, helping to tailor marketing strategies and predict sales.

1.7 Data Analysis vs. Statistics

Data analysis and statistics share many similarities, and their applications often overlap. However, they differ in their scope and specific functions. Data analysis encompasses a wider range of activities that are aimed at processing data and extracting useful information. In contrast, statistics is more narrowly focused on the collection, analysis, and interpretation of data, often emphasising inference and the quantification of uncertainty.

Generally (and somewhat superficially) **statistics** is a branch of mathematics that deals with collecting, analysing, interpreting, presenting, and organising data. As we mentioned the primary goal is to make inferences from a sample to a larger population. It focuses on the formulation of statistical models to understand underlying patterns and relationships. Traditionally, statistics has been used to draw conclusions about hypotheses and to estimate the reliability of hypotheses.

Data analysis involves processing and manipulating data with the goal of discovering useful information (somewhat similar to statistics). But its scope is generally broader, encompassing a variety of techniques to analyse data, which may or may not involve statistical methods. Data analysis can include data cleaning, transformation, and visualisation for example. It not only uses both descriptive and inferential statistical methods but also incorporates other techniques from data mining, machine learning, and big data analytics. The scope of data analysis is not limited to statistical studies. It also involves preparing data for analysis, cleaning it, and developing data-intensive products.

> *Example* 1.7.1: **Data Analysis**
>
> Consider a company that collects vast amounts of data about its customers, including their purchasing behaviours, preferences, demographics, and interactions with marketing campaigns. Rather than using traditional statistics to make inferences or test hypotheses about the data, the company applies machine learning algorithms to automatically group customers into segments based on similarities in their behaviours and characteristics.
>
> This approach is distinct from traditional statistics as it focusses on leveraging algorithms to discover patterns and make predictions, rather than testing a pre-existing hypothesis about the data. The use of machine learning allows for handling more complex datasets and drawing actionable insights, something which may not be straightforward or easy in statistical hypothesis testing.

Chapter 2
Types of Data

No data is clean, but most is useful.
– Dean Abbott, Co-founder at SmarterHQ

2.1 Qualitative (Categorical) Data

Qualitative or categorical data represent characteristics or attributes that cannot be measured on a numerical scale. Instead, they are categorised based on traits and descriptions. Qualitative data can be divided into two types: nominal data, which simply names or labels attributes without any order, and ordinal data, which involve some order or ranking of attributes.

> *Example* 2.1.1: **Qualitative (Categorical) Data**
>
> Qualitative or categorical data refer to data that can be divided into categories but that do not inherently have a numerical value. It is used to describe attributes or qualities of entities. Below are three examples.
>
> 1. **Eye Colour:** Categories could include blue, brown, green, etc. This type of data describes an attribute that does not have a natural numerical scale.
> 2. **Type of Cuisine:** This describes the kind of food served by restaurants, with categories such as Italian, Chinese, Mexican, Indian, etc.
> 3. **Marital Status:** This includes categories such as single, married, divorced, and widowed, which describe the legal relationship status of individuals.

> *Definition* 2.1.1: **Qualitative (Categorical) Data**
>
> **Categorical** data are observations that represent categories or labels. These data cannot be measured on a numerical scale but instead represent qualities or characteristics. Qualitative data can be further categorised into nominal and ordinal data types.

- **Nominal Data:** Nominal data are categorical data where the categories are unordered and do not have a natural or logical sequence. Examples of nominal data include gender (male, female), eye colour (blue, brown, and green), and types of vehicles (car, truck, and motorcycle).
- **Ordinal Data:** Ordinal data are categorical data where the categories have a natural order or ranking. However, the differences between categories are not necessarily uniform or measurable. Examples of ordinal data include ratings (e.g. Likert scale [7] ratings such as strongly agree, agree, neutral, disagree, and strongly disagree), educational levels (e.g. elementary school, middle school, high school, and college), and socio-economic status (e.g. low income, middle income, and high income).

2.2 Quantitative (Numerical) Data

Quantitative or numerical data represent amounts or quantities that can be measured on a numeric scale. This type of data can be further classified into two categories: **discrete data**, which consist of countable values or distinct whole numbers, and **continuous data**, which can take on any value within a given range and can be measured to any degree of precision.

> *Example* 2.2.1: **Quantitative (Numerical) Data**
>
> Quantitative or numerical data refer to data that can be measured and expressed numerically, allowing for arithmetic operations and statistical analysis. This type of data can be further classified into **discrete** data, which consist of countable values, and **continuous** data, which can take any value within a range. Below are some examples.
>
> 1. **Number of students in a class:** This is an example of discrete data, as you can count the exact number of students, such as 25 students.
> 2. **Height of individuals:** Heights are continuous data since they can be assumed any value, such as 1.75 m.
> 3. **Annual income:** This represents continuous data as well, though it is often rounded to the nearest unit of currency, like USD 50,000 per year.

Definition 2.2.1: **Quantitative (Numerical) Data**

Quantitative data, also known as numerical data, refers to information that can be measured and written down with numbers. These data are collected in the form of numbers or counts, with a numerical value that represents a measurement or a quantity. Quantitative data can be further classified into two main types

1. **Discrete Data:** This type of data represents items that can be counted and are listed as exact numbers. Discrete data are usually collected by counting, and they take on possible values that can be listed out. An example of discrete data is the number of students in a class.
2. **Continuous Data:** Continuous data represent measurements and therefore, unlike discrete data, can take on any value within a finite or infinite interval. This type of data is measured, not counted. An example of continuous data is the height of students in a class.

2.3 Level of Measurements

The levels of measurement refer to different ways in which variables or data can be categorised quantitatively and qualitatively. There are four primary levels of measurement: *nominal, ordinal, interval,* and *ratio.*

Definition 2.3.1: **Nominal Level of Measurement**

Nominal level of measurement refers to data that can be divided into categories that do not have a natural order or ranking.

For nominal level of measurement data can be classified into distinct categories, but there is no inherent order to these categories. Examples are gender (Male, Female, Other) or Blood Type (A, B, AB, O). Nominal data can be summarised for example using frequencies and percentages. You can use them for descriptive statistics like counts and mode.

Definition 2.3.2: **Ordinal Level of Measurement**

Ordinal level of measurement deals with data that can be placed in a natural order or sequence, but the intervals between the data points are not known or do not have a precise meaning.

Examples are educational level (Elementary, High School, College) or Satisfaction rating (Unsatisfied, Neutral, Satisfied). Ordinal data can be summarised for example using median and percentiles. You can use them in general for descriptive statistics.

> *Definition* 2.3.3: **Interval Level of Measurement**
>
> **Interval** level of measurement involves data that can be ordered, and the exact differences between the values are meaningful. However, there is no true zero point.

An example is calendar years. Interval data can be summarised for example using mean, standard deviation, and range.

> *Definition* 2.3.4: **Ratio Level of Measurement**
>
> **Ratio** level of measurement allows for the identification of the order, the exact value of the differences between the data points, and an absolute zero point.

Examples are weight, height, age, or income. Ratio data can be summarised for example using mean, standard deviation, and range. You can use them for parametric statistical tests and more advanced analyses, such as regression.

2.4 Cohort

The term *cohort* does not refer to a type of data but is often used in connection with datasets; it is important to know its exact meaning and how statisticians use it. In statistics, a **cohort** refers to a group of individuals (or objects) who share a common characteristic or experience within a defined period. Cohorts are often used in longitudinal studies (see below), in which researchers observe the same individuals over time to assess how specific factors affect them differently or to track changes and developments throughout their lives. A typical application is in the study of the effect of drugs, where researchers want to understand how a specific drug affects *similar* individuals.

> *Definition* 2.4.1: **Cohort**
>
> A **cohort** refers to a group of individuals who share a common characteristic or experience within a defined period.

The term "cohort" is particularly useful because it allows for the analysis of data that are similar between members at the beginning of the study, thus providing information on how particular conditions or experiences impact subjects over time.

Example 2.4.1: **Cohort**

A **cohort** could include all people born in a particular year and in a particular neighbourhood who are then studied over their lifetimes to investigate the influences of early life conditions on later health outcomes. Similarly, a school cohort might consist of all students who enter a university in the same academic year.

2.5 Longitudinal Data

Longitudinal data are collected over a longer period on the same subjects, allowing researchers to analyse and study changes over time.

Definition 2.5.1: **Longitudinal Data**

Longitudinal data are collected over a longer period on the same subjects, allowing researchers to analyse and study changes over time.

Example 2.5.1: **Longitudinal Data**

Consider this fictitious example: a study tracking the academic performance of a cohort of students throughout their educational life, from elementary school to high school. A group of researchers initiates a study to understand the impact of various teaching methods on student success over time. Starting in 2020, the researchers collect data **annually** on a specific group of students who begin the study as third graders. They gather information on each student's grades, test scores, participation in special programmes (such as tutoring or enrichment classes), and family history. The goal is to analyse how these factors contribute to student academic outcomes over the years. This study is **longitudinal** because it involves repeated observations of the same variables (e.g. academic performance) for the same subjects (the cohort of students) over an **extended period**. This allows the researchers to see how the students progress over time.

2.6 Cross-Sectional Data

Cross-sectional data are collected at a single point in time or over a very short period and provide a snapshot of a specific moment.

> *Definition* 2.6.1: **Cross-Sectional Data**
>
> **Cross-sectional data** are collected at a single point in time or over a very short period and provide a snapshot of a specific moment.

> *Example* 2.6.1: **Cross-Sectional Data**
>
> Consider a survey to assess the prevalence of obesity among adults in a city at a specific point in time. The survey is conducted, say, in December 2023. Data are collected from a large sample of adult residents in various neighbourhoods of the city. Each participant is asked, for example, about his or her height, weight, dietary habits, and exercise routines. The primary goal is to estimate the percentage of adults who are considered obese based on their Body Mass Index (BMI) and to identify any correlations between obesity rates and factors such as diet and physical activity. These data are collected at a single point in time (December 2023), rather than tracking the same individuals' health changes over multiple years. The data provide a **snapshot** of the health status of the community in December 2023.

2.7 Binary and Dichotomous Data

I would like to conclude this chapter with two data types that you might find in your statistical research: **binary** and **dichotomous**.

> *Definition* 2.7.1: **Binary Data**
>
> Binary data are data that have only two possible values (e.g. yes/no and on/off).

> *Definition* 2.7.2: **Dichotomous Data**
>
> Dichotomous data are a special case of binary data where the two values are mutually exclusive (e.g. male/female).

2.7 Binary and Dichotomous Data

Example 2.7.1: **Non-dichotomous vs. Binary Data**

Suppose a survey asks respondents whether they enjoy a specific type of music, with the options being "Yes" or "No". Here, the data are **binary**, since there are only two response options.

However, this set-up could be **non-dichotomous** if the question does not cover all possibilities or its outcomes are not complete opposites. For example, if the survey question is about a preference ("Do you like classical music?"), and the possible answers are "Yes" and "No", there are inherent limitations. The "No" answer could imply different things, maybe the person answering dislikes classical music, is indifferent to it, or is unfamiliar with it. The binary response does not capture these subtle differences. There is a possibility that someone may like more than one type of music, which is not exclusively captured by a simple yes/no to one genre. In this case, the binary data are not dichotomous because "No" does not strictly mean the opposite of "Yes" (i.e. dislike as opposed to merely lacking a preference or being unfamiliar).

You need to ensure that the data adequately represent the phenomena being studied and that any conclusions drawn are valid given the potential ambiguities in the way responses are categorised (or the questions are asked). Thus, while binary data always involve two categories, it is the nature of these categories and their relationship to each other that determine whether the data are truly dichotomous.

Chapter 3
Data Collection Methods (Sampling Theory)

3.1 Introduction

Think of sampling theory as a set of techniques that help you choose the best pie slices that give you a real taste of the whole thing. This helps you to make solid guesses and to build or test statistical hypotheses that actually make sense according to what you need to find out.

Often, when performing statistics, you will find yourself using datasets that have been previously prepared and curated by others. However, when starting a new research project, you may be faced with the task of creating a completely new dataset that will be used to answer a specific research question. Acquiring data are trickier than you can imagine, and knowing the fundamentals of sampling will help you assess bias, representativeness, and much more about your data. It will crucially help you in deciding **what** kind, **how much**, and **how many types** of samples to acquire for a given research question. Let us consider a couple of examples on how acquiring data may not be as easy as you may think.

- Conducting surveys to collect data on homeless populations poses significant challenges due to the often hidden nature of this group. The difficulties arise in various stages of the data collection process. Obtaining a representative sample is difficult. Surveys conducted in shelters or food distribution centres may not capture those who avoid these services. Furthermore, the timing and location of data collection can significantly affect who is included in the sample. The response rates among homeless populations are also generally low due to mistrust, mental health issues, or lack of interest.
- A longitudinal study (see Sect. 2.5) aims to collect data over extended periods to observe the development and progression of chronic diseases such as diabetes or heart disease. This type of study design faces several challenges. Recruiting a large and diverse sample that is willing to participate over many years can be difficult. High attrition rates are common, as participants may lose interest, move away, or die during the study period. Maintaining consistency in data collection methods over time is crucial, but challenging, especially with changes in technol-

ogy and staff turnover. Ensuring that participants follow the same protocols for self-reported data, such as diet or physical activity, requires on-going education and support. Self-reported data might also be unreliable. Regular medical exams and tests must be standardised to ensure comparability of data over time (something that may be overly difficult over multiple years). In addition, long-term data collection raises concerns about data privacy and ethical handling of sensitive health information. Obtaining and renewing informed consent over many years adds complexity to the study.

These two simple examples should give you at least an intuitive idea of how many different challenges you must face when defining what new data to acquire and collect for a statistical project. In this chapter, we review the fundamental concepts of sampling that will help you design a proper data acquisition strategy for your projects.

3.2 Research Questions and Hypotheses

We now discuss two important concepts that drive the creation of datasets: research questions (RQ) and hypotheses. Research projects often start with a dataset. Someone finds or is given a dataset and is asked to analyse something. Sounds familiar? Maybe the professor you are working with got a dataset from some other research group and would like you to *try something with statistics or machine learning* or ask you directly to do something specific.

This is how most of these projects begin. If you find yourself in this situation, take a step back and think about the problem you are trying to solve. What questions are you trying to answer? What hypotheses do you have? It is important to clarify some terminology and explain what a research question or hypothesis is, why you need one, and how designing a good hypothesis to verify (or disprove) is mandatory for good dataset creation. Before we get to hypotheses, let us start by defining what a research question is.

3.2.1 Research Questions

A good research project starts **always** with one (or multiple) research question (RQ). We can loosely define it as a concise, focused inquiry formulated to address a specific concern or knowledge gap within a broader topic area (this is a very convoluted definition, right?). You will find some examples below that should help you get an intuitive idea of what a RQ is.

3.2 Research Questions and Hypotheses

> *Definition* 3.2.1: **Research Question**
>
> A research question is a concise and focused inquiry formulated to address a specific concern or knowledge gap within a broader topic area.

Since this definition is quite generic (and a bit pompous), let us give three examples to help you understand the concept.

- How does the introduction of non-native plant species affect biodiversity in urban green spaces?
- Can we predict the onset of diabetes from the medical history of patients?
- What impact does the integration of technology in the classroom have on student engagement and learning outcomes?

By looking at the examples, you will realise that the questions are general in nature and cannot really be answered with some precise number (take the last question, what does *impact* mean, and how can it be measured?). They guide the research process, determine the direction of the study, and can give some hints on how to decide what types of data or methods need to be collected and used. In any research project, you should start with a research question (or multiple ones, if relevant). This is the typical approach that you use if you are working on your thesis (Bachelor, Master, or even Ph.D.) by the way. A good RQ gives the context of the research project, hints at its impact, and highlights its importance. It should be understood by non-experts and inspire. In addition, it typically allows for a wide range of outcomes.

As its name implies, it is good practice to formulate it as a **question**. Avoid statements that are not formulated as questions.

3.2.2 Hypothesis

Once you formulate your RQ, you will need hypotheses. Whether you can disprove or verify.[1] Hypotheses can be loosely defined as a prediction of the relationship between two or more variables. It can be described as an educated guess about what happens in an experiment. Researchers tend to use hypotheses when significant knowledge on the subject is already available. After the hypothesis is developed, the researcher can develop or gather data, analyse them, and use them to support or negate the hypothesis.

---

[1] Karl Popper, the famous philosopher, would disagree on the verification of a scientific hypothesis, but we will skip this discussion here.

> *Definition* 3.2.2: **Hypothesis**
>
> A hypothesis is a prediction of the relationship between two or more variables. It can be described as an educated guess about what happens in an experiment. Researchers tend to use hypotheses when significant knowledge on the subject is already available.

Some examples are as follows.

- Global warming has increased sea level by 1 cm in the past 10 years on average worldwide.
- The number of cars on the road each day on average in Zürich has decreased by 5% after the COVID year.

You should immediately see the difference between a hypothesis and a research question. While an RQ is written as a question (hence the name), a hypothesis is always written as a statement that can be verified or disproved. A hypothesis is the fundamental building block that allows you to design experiments to test the hypothesis itself. In other words, it means that getting the right data is a consequence of a well-thought hypothesis.

To summarise what we discussed when starting a new research project, you should proceed according to the following steps:

1. Formulate one (or multiple, but not more than 2–3) research question.
2. Formulate a series of hypotheses that will help you answer your research questions.
3. Design experiments to verify or disprove the hypotheses formulated.

During the work of points 2 and 3 you will find that you have enough information to be able to design your data collection strategy.

3.3 Survey Sampling

Generally speaking, statistics is the science of drawing statements and conclusions about a population by using a sample of it (as we discussed earlier). Let us summarise some terminology here again so that you do not have to jump back and forth between sections.

The term **population** normally refers to a set of objects (patients, molecules, galaxies, etc.) that are *infinite* (at least in theory) in nature and, due to this, cannot be known or described exactly. For example, the population of all results of tossing a coin is an infinite set of two possible results: head and tail. We do not have access to the infinite set, and thus we try, with statistics, to study and draw conclusions about it from a finite set of results. In reality there are no sets of objects that are infinite (e.g. all persons below 18 years of age who have lived so far on Earth) but are large enough that it is impossible to know or describe. So, when defining populations, you can substitute the word *infinite* with the words *very large*.

3.3 Survey Sampling

We can now give an intuitive definition of survey sampling (or simply sampling) now. Survey sampling (or simply sampling) deals with the problem of selecting a finite set of elements from a potential infinite population.

Definition 3.3.1: **Survey Sampling**

Survey sampling (or simply sampling) encompasses a set of techniques that deal with the problem of selecting a finite set of elements from a potential infinite population.

The term survey refers to the act of collecting data.

Tip 3.3.1: **Meaning of Survey**

The word **survey**, according to the Britannica dictionary, refers to *an activity in which many people are asked a question or a series of questions to gather information about what most people do or think about something*. But the dictionary also gives the definition of *an act of studying something in order to make a judgement about it*. This second definition is much more apt and will serve the reader well. Practically, all data used in the scientific field are not coming from surveys but from measurements and experiments, and thus, I much prefer the second definition.

In our initial discussion, you may have got the impression that populations are given and simply exist. But even **defining** if an object is part of a population is not trivial. For example, suppose that you want to include in your population all smokers. How do you define if someone is a *smoker*? For example, a possible definition is [8] "*an adult who has smoked more than 100 cigarettes in his/her lifetime and currently smokes at least once a week*". Assessing whether a person is a smoker may not be as easy as it sounds. In sampling theory, you should always define what are called **eligibility criteria** that will define which object is part of the population and which not. This step is fundamental, but its relevance is particularly evident in medicine when selecting which patients should be in the population and which are not depend strongly on the specific RQs in the medical context.

After designing your RQs and hypotheses, the next step is to define your population. In other words, you must design eligibility criteria that would define your **hypothetical** population. I have used the term hypothetical, meaning that you do not yet have data on all individual units in the population. Having the criteria allows you to decide whether, when presented with an object, you can decide if this belongs to your population or not.

Now we need to discuss two important concepts to create a sample from a population: **probability** and **non-probability** sampling.

3.3.1 Non-probability Sampling

Non-probability sampling simply means selecting elements from the survey population according to fixed rules and not by chance. Sometimes, a specific sampling strategy is chosen due to specific limitations (such as time or budget constraints).

> **Definition 3.3.2: Non-probability Sampling**
>
> **Non-probability sampling** simply means selecting elements from the survey population according to fixed rules and not by chance.

Here are some examples of non-probability sampling.

- **Restricted sampling**: Sampling is simply done keeping only parts of the population that are easily accessible. Maybe you are working in a hospital, and thus your sample includes only patients from your hospital.
- **Judgement sampling**: Sampling is obtained based on what the sampler believes to be *representative*. Maybe you are studying brain tumours, and how they appear in MRI images. You may decide to study only specific types of tumour, as your experience has shown that, in general, they appear in MRI images similarly to most tumours.
- **Convenience sampling**: Sampling is performed simply by keeping what is easily *reachable*. This type refers more to classical surveys, in which people had to reach people to ask questions.
- **Quota sampling**: The sample is gathered by several interviewers (e.g. when talking about surveys), each tasked with collecting a specific quantity of units that possess particular types or characteristics. The selection of these units is entirely up to the discretion of the interviewers. If you are not dealing with interviews, you may have a certain number of people, each tasked with getting a certain number of objects you want to study (e.g. you may have a certain number of chemists, each tasked with getting a certain number of chemicals for your study).

These methods are used when sampling by chance (see the next section) is not feasible or simply too time-consuming or expensive. Statistical validity of the results with such samples relies **strongly** on assumptions. For example, in judgement sampling, the analysis relies on the assumption that the sample is representative, something not everyone may agree on. Typically, these methods are chosen because they are faster or less expensive. Non-probability sampling is a widely used method for gathering essential data on human populations, extensively employed by researchers in social and behavioural sciences, as well as those in medical and health fields. The volume of data required can be vast, encompassing tens or even hundreds of measurements per participant (blood samples, urine analysis, medical imaging, psychological assessments, etc.). For such studies, non-probability sampling might represent the only practical method.

3.3 Survey Sampling

The process works according to the following steps.

1. **Define the population**: Determine who or what you want to study. Unlike probability sampling (more on that later), you do not need to have a complete list of the population (in other words, you do not need, e.g. a list of names of all possible participants in your study). You just need criteria on how to define your population.
2. **Choose a non-probability sampling method**: Select the most appropriate non-probability sampling technique based on your RQs and hypotheses and the nature of your population. Common methods we mentioned include convenience sampling, judgement sampling, etc.
3. **Determine sample size**: Decide on the number of participants or objects you need. This decision may be influenced by factors such as the depth of analysis required, the time, and the available resources (it may very well be a budget issue). Non-probability sampling does not rely on statistical formulas for sample size determination and is typically determined by practical reasoning (like how much money you have, how much time, etc.).
4. **Recruit participants**: Based on the chosen method, begin recruiting participants or getting your objects. For example, in convenience sampling, you collect data from individuals who are readily available. If you are studying chemical compounds, you would need to go and buy them for your study. You may decide to buy only compounds that are safe to use, or inexpensive to buy, for example.
5. **Collect data**: Once your sample is selected, collect the data necessary for your study. This could involve surveys, interviews, observations, etc.

Remember, while non-probability sampling can be more practical or the only option available for certain studies, it may introduce bias and limit the generalisability of the results.

3.3.2 Probability Sampling

Probability sampling simply means that a sample is obtained by selecting elements of the population in a random fashion (more precisely according to some probability measure). Each element is given a probability of being selected (an easy approach is to give all elements of the population the same probability of being selected) to remove bias associated with subjective decisions (e.g. if using judgement sampling). If you have a large population at your disposal, you can use this approach to randomly select a sample.

> *Definition* 3.3.3: **Probability Sampling**
>
> **Probability sampling** means that a sample is obtained by selecting elements of the population in a random fashion (more precisely according to some probability measure).

The process works according to the following steps.

1. **Define the population**: Determine the entire group of individuals you want to study. This could be all students in a school, all employees in a company, etc. This comes from your RQ and hypotheses, as we discussed.
2. **Create a list**: Compile a complete list of all members of the population. Each member is assigned a unique identifier, such as a number or string.
3. **Random selection**: Use a random method, such as a random number generator, to select a specific number of individuals/objects from the list you created in the previous step. The number of individuals selected depends on the desired/possible sample size. Note that depending on your RQ or hypotheses, not all elements of the population must have the same probability of being chosen.
4. **Conduct the survey**: Gather data from individuals or objects chosen by random selection. The sample size determined in the previous step is directly influenced by the data requirements. For example, if acquiring the information is costly, you might opt for a smaller sample size.

The random method you use in step 3 can be more complicated than using an equal probability for each individual/object in your population. That will depend on your RQ and the hypothesis you are studying. Note that in this case, you need to have a list of all individuals (or objects) belonging to the population. So it is not enough to have eligibility criteria. This might make data collection a more challenging task.

3.4 Stratification and Clustering

A population is stratified if it is divided into $q \in \mathbb{N}$ non-overlapping groups (called **strata**).

> *Definition* 3.4.1: **Stratified Population**
>
> A population is stratified if it is divided into q non-overlapping groups (called **strata**).

For example, patients can be divided into subgroups each having a different disease, into different age groups, etc. A population is said to be clustered if it can be divided into subgroups (called **clusters**).

Definition 3.4.2: **Clustered Population**

A population is said to be clustered if it can be divided into subgroups (called **clusters**).

The two definitions may seem exactly equivalent, but the difference lies in the way in which *clusters* and *strata* are used. When dealing with a stratified population, sampling involves selecting elements from **all** strata. When dealing with a clustered population, only a portion of the clusters will appear in the final sample. For example, if we stratify patients in different age groups, then when creating our sample we will select patients from **every** age group in the population. On the contrary, if we cluster people geographically, our final sample may contain only a subset of regions that we have at our disposal.

3.5 Random Sampling Without Replacement

In this method, once an individual or object is selected from the population for inclusion in the sample you are creating, it cannot be chosen again. This approach ensures that each member of the population can be selected, but no individual or object can be included in the sample more than once. It is typically used when the goal is to avoid duplicating members in the sample, since your RQ or hypotheses require it. For example, if you are drawing cards from a deck, once a card is drawn, it is set aside and not put back into the deck for subsequent draws.

Warning 3.5.1: **Random Sampling Without Replacement**

Here is an overview of the advantages and disadvantages.

- **Advantages:** The method ensures each member of the population can be selected only once, preventing duplicates in the sample. This can lead to a more diverse and representative sample. Additionally, the maximum sample size is limited to the population size, making it easier to manage.
- **Disadvantages**: The method requires a comprehensive list of the population beforehand, which can be difficult or impractical to obtain for large or dynamic populations. In addition, managing and tracking selections from a very large population can be more complex and resource intensive.

> *Definition* 3.5.1: **Random Sampling Without Replacement**
>
> This is a sampling technique where once an individual or object is selected from the population for inclusion in the sample you are creating, it cannot be chosen again. This approach ensures that each member of the population can be selected, but no individual or object can be included in the sample more than once.

3.6 Random Sampling with Replacement

In this method, after an individual or object is selected from the population and included in the sample, it is re-inserted into the population, making it eligible for selection again in the next iteration. This method allows for the possibility of the same individual being put multiple times in the sample. It is particularly useful in simulations and bootstrap methods (we will only briefly discuss bootstrap, as it goes beyond the scope of the book), where the objective is to create multiple independent samples from a single dataset to estimate the distribution of a parameter. An example of this would be to draw a card from a deck, noting its value and then putting it back in the deck before drawing again. Note that with this approach, you can create a sample that is larger than the population from which you are sampling, since you can select an element multiple times. You should be very careful to do so to avoid introducing bias in your sample that could skew your statistical results.

> *Warning* 3.6.1: **Random Sampling with Replacement**
>
> Here is an overview of the advantages and disadvantages.
> - **Advantages:** Each selection is independent of the others, making the process simpler and often more suitable for theoretical or computational studies, like bootstrapping. Furthermore, the method allows for a sample size that can exceed the population size, providing flexibility in experimental design and analysis. It can be particularly useful when the population size is small, as it allows a larger sample size without the constraint of exhausting the population.
> - **Disadvantages**: When using this method there is the possibility of selecting the same individual or object multiple times, which can lead to duplicates in the sample, affecting diversity and potentially skewing results. Furthermore, the presence of duplicates might result in biased estimates of population parameters if not properly accounted for in the analysis. Finally, while it allows for greater sample sizes, it may result in a sample that is less representative of the population, especially if the population is large and diverse.

Definition 3.6.1: **Random Sampling with Replacement**

This is a sampling technique where after selecting an individual or object from the population and including it in the sample, it is placed back into the population, making it eligible for selection again in the next iteration.

3.7 Random Stratified Sampling

If you have a stratified population, you should pay attention to having all strata in your sample. You can obtain this by using the following process, assuming that you have your population, but you have not yet stratified it. The following steps will help you obtain a stratified population and sample.

1. **Identify the stratifying variable**: Choose the features you will use to divide the population into different strata and decide how to split your population in strata. This should be a characteristic that is believed to influence the outcome of the research, such as age, sex, income level, etc. In addition to deciding, for example, that age should be used for stratification, you must also decide in which age groups you want to stratify your population. This is a two-step process: Decide on the features and the ranges of features that will define your strata.
2. **Divide the population into strata**: Based on the stratification strategy from the previous point, divide the population into distinct strata. Each unit in the population should belong to one and **only** one stratum.
3. **Determine sample size**: Decide on the total sample size for your study.
4. **Define the sample size for each stratum**: Determine how many individuals/objects to sample from each stratum. This can be done proportionally (*proportional allocation*) based on the size of the strata in the population or equally among the strata regardless of their size in the population or based on other considerations relevant to your RQs or hypotheses.
5. **Select samples from each stratum**: Within each stratum, use random sampling to select individuals. This ensures that every member of the stratum has an equal chance of being included in the sample. You will select as many samples from each stratum as defined in the previous step.
6. **Collect data**: Proceed to collect data from selected individuals in all strata.

By following these steps, stratified random sampling allows you to obtain a sample that is more representative of the population (meaning that the statistical results obtained from the sample should reflect the characteristics of the population), especially when there are significant differences between strata that could affect the study's outcomes. Consider an example to make this process and reasoning more concrete. Imagine a scenario in which you are trying to assess the impact of a new teaching method on student performance in mathematics across a region. This region includes a diverse array of schools, such as public and private, as well as urban and rural, each with varying levels of resources and socio-economic backgrounds among their students. The socio-economic status of students is a significant factor that can influence academic performance, with schools in rich areas typically having more resources and, consequently, potentially better student outcomes compared to schools in less wealthy areas.

To directly compare average scores across the entire region without accounting for these socio-economic differences could obscure the true effectiveness of the teaching method. This is because the method could perform differently in different environments: being more effective in some and less effective in others. Stratified sampling addresses this issue by ensuring that schools from each socio-economic category are represented in the sample. This allows for a more precise analysis of the teaching method's effectiveness across diverse socio-economic backgrounds.

In this context, you would need to first categorise the schools into different strata based on their socio-economic status (e.g. high, medium, and low). Then you would need to decide on the sample size for each stratum to ensure proportional representation based on the number of schools or the student population within each socio-economic category. After this, schools and, subsequently, students within those schools are randomly selected from each stratum to participate in the study. The data on student performance in mathematics are then collected and analysed, with the analysis making comparisons both within and across the different socio-economic strata to assess the overall impact of the teaching method. Stratified sampling as described enables you to draw more accurate and generalisable conclusions regarding the effectiveness of the teaching method.

3.8 Bootstrap

I would like to mention a last important statistical resampling method that is widely used: bootstrap. It was developed originally by Efron in 1979 [9] and has been extensively studied by many statisticians. I will not go into many details, as there are many books that do that already. If the reader is interested, the book by Michael R. Chernik *Bootstrap Methods A Guide for Practitioners and Researcher* is a very complete introduction to the subject [10]. Its 200 pages of references (no, is not a typo) gives a good idea about the amount of research done in this area. Another self-contained graduate text that can be consulted is the one by Davidson and Hinkley from 1997 [11]).

In statistics, bootstrap refers to a resampling technique used to estimate the distribution of a statistic (like the mean, median, or variance, for example) by sampling with replacement from the original data. It is particularly useful when it is difficult to make assumptions about the population or when the sample size is small. This method allows statisticians to make inferences about the population from which the sample is drawn without relying on traditional parametric assumptions. The method in its most basic form is quite easy to understand. It works in the following way:

3.8 Bootstrap

1. **Original Sample**: You start with an original dataset of size n.
2. **Perform Resampling**: Generate m new datasets (called bootstrap samples) by randomly selecting n observations with replacement from the original dataset. Each bootstrap sample will be the same size as the original data but may contain duplicate observations.
3. **Evaluation of statistic estimator of interest**: For each bootstrap sample, compute the statistic of interest (e.g. mean, median, variance, etc.). With m large enough, you will have a distribution of the statistic based on these resamples. This distribution is called the bootstrap distribution.
4. **Evaluate confidence intervals and distribution characteristics**: Use the bootstrap distribution to estimate confidence intervals, standard errors, or p-values for the statistic of interest.

Bootstrap is often used when it is not possible to evaluate things like confidence intervals by theoretical means. Its statistical underpinnings and its justification go well beyond this book but it is a very versatile technique that can be very useful in dealing with complex and difficult to evaluate statistical estimators. There are of course downsides to using bootstrap, the major one being its computationally intensive nature. You will have to calculate your statistics m times, and if m is large enough, that may be very computationally intensive. Also care must be taken when applying it to very small samples as this may turn out to be tricky in the interpretation.

Let us show a toy example to make this discussion more comprehensible. Let us consider the array $a = \{2, 8, 5, 9, 12, 7, 6, 11, 4, 10\}$. Suppose that you are interested in estimating the mean of the population of which a is a sample. If you calculate the mean of a (indicated here with $\bar{a}$), you will get

$$\bar{a} = 7.4 \qquad (3.1)$$

but suppose instead of the 10 values, you have only 9. Suppose that you consider a subset of a, such as $b = \{2, 8, 5, 9, 12, 7, 6, 4, 10\}$. In this case the average of b (indicated with $\bar{b}$) is $\bar{b} = 6.3$. So how sure can you be that your initial estimate is correct? Using bootstrap as described in this section, you can easily calculate the confidence intervals of $\bar{a}$ by using bootstrap samples. Your bootstrap samples may look like $\{2, 8, 8, 9, 12, 7, 4, 4, 10\}$ or $\{2, 2, 5, 9, 12, 7, 4, 4, 10\}$ for example. Remember a bootstrap sample of a is obtained by sampling ten values from a **with** repetitions.

In Fig. 3.1 you can see the distribution of values of $\bar{a}$ and the 2.5th and 97.5th percentiles (do not worry now about what percentiles are, as I will explain them in Chap. 5) obtained with 10^5 bootstrap samples.

As any method, bootstrap has limitations. Let us discuss some of them. To start with, care must be taken when considering small samples. If your original sample size (in our example a) is very small, bootstrap resampling may not provide reliable estimates. With small datasets, there may not be enough variability in the resamples, leading to biased or overly optimistic estimates of confidence intervals or other statistics. We have not discussed this, but bootstrap assumes that the data points are independent of each other. If the data are correlated (e.g. time series data, spatial data, or repeated measures from the same subject), standard bootstrap resampling may

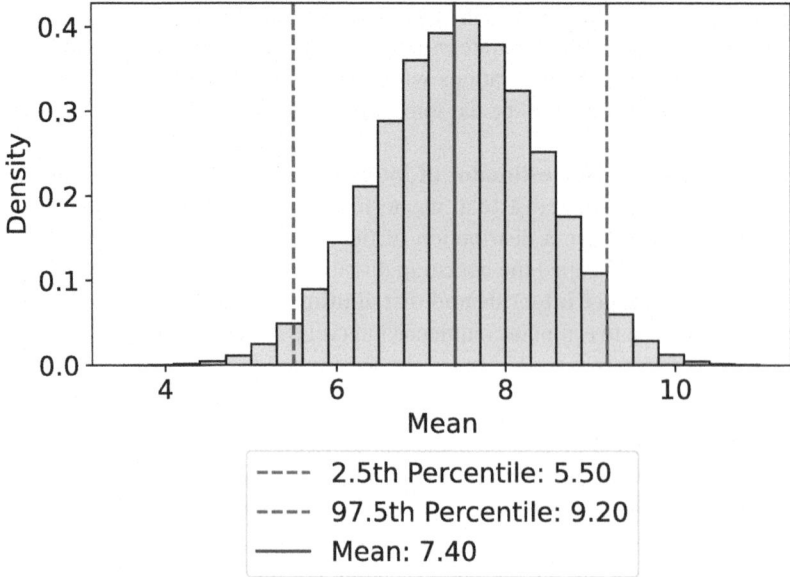

Fig. 3.1: The distribution of the values of the mean of the array $a = \{2, 8, 5, 9, 12, 7, 6, 11, 4, 10\}$ obtained with 10^5 bootstrap samples. The blue line indicates the average and the red dashed lines the 2.5th and 97.5th percentiles

lead to incorrect estimates of variability. For dependent data, specialised techniques like the block bootstrap should be considered (for a treatment of the subject, you may want to check Chapter 5 in Chernick's book [10]).

Outliers must also be treated with care when using bootstrap. Since resampling is done with replacement, outliers may be over-represented in some resamples, distorting estimates such as mean or confidence intervals. As such, bootstrap can magnify the impact of outliers, leading to biased results. The choice of statistic is also important. Bootstrap works best for some statistics (e.g. means and medians) but may not be appropriate for all statistics. For example, estimating very rare events or extreme percentiles (e.g. the 99th percentile) may not perform well because the original sample may not capture enough instances of those rare events. You should also be aware that estimation of statistical estimators with bootstrap does not automatically correct for bias. If your original estimator is biased or misleading (e.g. the sample mean of a skewed distribution), the bootstrap resamples will also reflect that. More fundamentally than any other limitation, bootstrap assumes that the original sample is **representative** of the population. If the original sample is biased or not representative of the population, the bootstrap can amplify those issues, leading to incorrect inferences. It is crucial to consider the quality of your original data before applying the bootstrap (that is anyway a good idea generally speaking, if your data are bad, your statistical analysis will also be bad).

3.8 Bootstrap

As a last warning, bootstrap confidence intervals should be interpreted with care. Different methods for calculating these intervals (e.g. percentile method, bias-corrected and accelerated (BCa) method, described in [10] as *accelerated constant* in case you want to look it up) may yield different results.

This small section has only the goal of making the reader aware of bootstrap, its complexity, and limitations and has no presumption whatsoever of giving a proper treatment of the subject. The study of bootstrap is somewhat more advanced in nature and does not fit this short book. Nonetheless the reader is advised to study it and master it, as this is a widely used technique in statistics.

Chapter 4
Measures of Central Tendency

4.1 Introduction

Measures of central tendency are statistical tools used to summarise a set of data by identifying the central point around which all other data points cluster. The main measures are **mean, median, mode,** and **mid-range**. To define them, consider a set of measurements x_i of a random variable X (to review what a random variable is, see Chap. 1), where i indicates the index of your measurements. Suppose also that we have N measurements in total. In the next sections, we will discuss mean, median, and mid-range in detail.

4.2 Mean

The **mean** of a dataset composed by elements denoted by x_i is defined by

$$\text{mean of } X = \frac{1}{N} \sum_{i=1}^{N} x_i \tag{4.1}$$

where the symbol $\sum$ indicates summing all elements x_i for i varying from 1 to N. In other words

$$\text{mean of } X = \frac{1}{N} \sum_{i=1}^{N} x_i = x_1 + x_2 + \ldots + x_N \tag{4.2}$$

The mean is often indicated with $<X>$ or $\bar{X}$. Often, in statistics, we talk about the **expected value** of a random variable X.

> *Definition* 4.2.1: **Mean (Average)**
>
> $$\text{mean of } X \equiv \bar{X} = \frac{1}{N} \sum_{i=1}^{N} x_i \qquad (4.3)$$
>
> where the symbol $\sum$ indicates summing all elements x_i for i varying from 1 to N.

This is indicated by the symbol

$$\mathbb{E}(X) \qquad (4.4)$$

and for all practical purposes it can be thought as the mean of the random variable X. A more nuanced discussion can be found in the next *info* box.

> *Tip* 4.2.1: **Expected Value**
>
> The term **expected value** in statistics is rooted in the concept of predicting the average outcome of a random variable over a long series of experiments. It is called "expected" not because it is what will definitely happen in a single event or a few events, but because it represents the average or "expected" outcome over a vast number of trials.
>
> Here is an intuitive way to understand it: Imagine you are flipping a fair coin. The expected value of the number of heads in one flip is 0.5. This does not mean that you expect to see half a head when you flip a coin once. Instead, it means that if you were to flip the coin many, many times, the average number of heads per flip you would expect to see is 0.5. In other words, about half of the flips would be heads, and the other half would be tails, over a long series of flips.
>
> In a more formal mathematical sense (and in the language of outcome spaces and events), the expected value is calculated by multiplying each possible event by its probability of occurrence and then summing all those products. It is a fundamental concept, especially in probability theory and statistics, because it provides a single summary number, an "expectation", of what you predict the average outcome to be if you were to repeat a random experiment an infinite number of times.
>
> So, the expected value is a kind of long-term average that might never be actually observed in any single experiment but is what you would expect to happen on average if the conditions were repeated over and over again.

You should **not** use the mean as a measure of central tendency in the following situations. When the data distribution is skewed, in other words, asymmetric (discussion of skewness can be found in Sect. 9.2), the mean can be significantly affected by the extreme values (outliers, discussed in Chap. 7), leading to a misleading representation of the central location of the data. In such cases, the median is often a better measure because it is not influenced by outliers and more accurately reflects

4.2 Mean

the centre of a skewed distribution (more about the median in Sect. 4.3). Note that outliers can strongly influence the mean, pulling it towards the extreme values and away from the typical value of the majority of the data points. If your data include outliers that you cannot remove or adjust for, using the median (see Sect. 4.3) is generally more appropriate. Categorical data represent categories with a meaningful order, but they are not continuous measures of some characteristics (blue eye is not necessarily larger or smaller than brown eyes). Since the mean involves arithmetic operations, it is not applicable to categorical data. The mode discussed in Sect. 4.4, being a positional measure, should be used instead.

There is another case that you should be aware of. For discrete variables that assume only a limited number of integer values (consider throwing a dice), the mean might result in a value that is not a possible outcome of the data. For example, the average number of children per family in a survey might be 2.4, but families cannot have 0.4 of a child. In such cases, the median or mode might provide a more useful measure of the central tendency, depending on its planned use. While the mean is a widely used measure of central tendency because of its mathematical properties and relevance in various statistical analyses, it is crucial to assess the nature of your data and the impact of outliers, the data distribution, and the scale of measurement before deciding whether the mean is the most appropriate measure to use.

Example 4.2.1: **Mean of an Array of Numbers**

Let us calculate the mean age of a group of individuals that have the following ages (measured in years): 23, 29, 31, 35, 22, 27, 30, 26. This can be done easily by following the next steps:

1. Sum all the ages in the dataset: $23 + 29 + 31 + 35 + 22 + 27 + 30 + 26 = 223$.
2. Count the number of individuals: 8.
3. Divide the total sum of ages by the number of individuals to find the mean:
$$\text{Mean} = \frac{223}{8} = 27.875$$

Now imagine we add to the group 2 people who are 99 years old (those would be classified as outliers in this case). Suddenly, the mean jumps to 42.1 years. This is clearly not representative of the majority of the ages (which goes from 23 to 35). So care must be taken in using the mean, depending on the number and importance of the outliers.

Outliers are an important aspect in statistics and will be discussed and defined later in this book in Chap. 7 since to define them an understanding of the measures of dispersion is necessary.

4.3 Median

The median of a dataset is the value that separates the upper half from the lower half of the data. To calculate the median, you follow this algorithm.

1. Sort the data in non-decreasing order so that now $x_1 \leq x_2 \leq x_3 \leq \ldots \leq x_N$.
2. Determine the middle position(s):
   - If the number of data points, N, is odd, the middle position is
   $$\frac{N+1}{2}$$
   - If N is even, the two middle positions are
   $$\frac{N}{2}$$
   and
   $$\frac{N}{2}+1$$
3. Calculate the median:
   - If N is odd, the median is the value at the middle position $x_{(N+1)/2}$.
   - If N is even, the median is the average of the values at the two middle positions
   $$\text{median} = \frac{x_{N/2} + x_{N/2+1}}{2}$$

> *Definition* 4.3.1: **Median**
>
> The median of a dataset is the value that separates the upper half from the lower half of the data.

Although the median is a robust measure of central tendency that is less affected by extreme values (outliers) compared to the mean, there are certain situations where using the median is not the best choice. The following are some of those cases.

1. **Data with a uniform distribution:** For data that are uniformly distributed, where all values occur with the same frequency, the median does not provide more information than the mean, and using the mean might give a more **intuitive** sense of "average".
2. **Highly skewed data with large outliers:** In cases where outliers represent important and meaningful information (e.g. income distribution with extremely high incomes), focussing solely on the median will ignore significant aspects of the data distribution.

3. **Nominal data:** The median requires that the data can be ordered. For nominal data, which consist of names or categories without an inherent order, the median is not applicable, and mode is used instead.
4. **When detailed distribution information is needed:** The median provides a midpoint of the data but does not offer insights into the distribution's shape or spread as the mean might when combined with measures like standard deviation.

> *Example* 4.3.1: **Median of an Array of Numbers**
>
> The median is a measure of central tendency that identifies the middle value in a dataset when it is ordered from smallest to largest. Consider the following set of numbers: 2, 3, 5, 7, 11. To find the median we need to do the following.
>
> 1. First, we arrange the numbers in ascending order. In this case, our set is already ordered: 2, 3, 5, 7, 11.
> 2. Since there are five numbers, an odd quantity, the median is the middle number. This makes our calculation straightforward.
> 3. The median is the third number in the ordered set, which is **5**.
>
> Now, consider a set with an even number of observations: 2, 3, 5, 7. In this case the median can be calculated with the following steps:
>
> 1. The set is already in ascending order: 2, 3, 5, 7.
> 2. With four numbers, we take the average of the two middle numbers, 3 and 5.
> 3. The median is $(3 + 5)/2 = 4$.
>
> The median provides a valuable measure of the centre of a dataset, especially useful for skewed distributions or when outliers are present.

4.4 Mode

The mode of a dataset is the value or values that appear most frequently. It is a measure of central tendency that is particularly useful for categorical data. Unlike the mean and the median, the mode can be applied to data of any type: numerical, categorical, or ordinal. A dataset may have one mode (then we speak of the dataset to be unimodal), more than one mode (bimodal or multimodal), or no mode at all if no value repeats itself.

Finding the mode involves identifying the value or values that occur most frequently in the dataset. The following steps outline a basic algorithm for finding the mode:

1. Count how many times each value appears in the dataset.
2. Identify the value or values with the highest count (thus finding the **mode**).
3. If no value appears more than once, the dataset does not have a mode.

The mode is particularly useful in the following scenarios. For data that cannot be quantitatively measured but can be categorised, the mode identifies the most common category, thus giving information on the element in the dataset appearing more often. Furthermore, in distributions with multiple peaks (or multiple elements appearing often), the mode can help identify different groups within the data that might be masked by measures like the mean or median.

> *Definition* 4.4.1: **Mode**
>
> The **mode** of a dataset is the value or values (in this case we talk about **modes**) that appear most frequently. A dataset may have one mode (that we speak of the dataset to be **unimodal**), more than one mode (**bimodal** or **multimodal**), or no mode at all if no value repeats itself.

The mode should not be used (or used carefully) in the following scenarios. In datasets with more than one mode (multimodal), the mode can become less informative, as multiple values are equally common. For datasets where values are evenly distributed or occur infrequently, the mode might not exist or may not provide meaningful information (what about a dataset where each element appears only twice?). Furthermore, the detection of the mode is highly sensitive to sample size. Small changes in data can lead to different modes, making it potentially unstable in small datasets or those with a lot of unique values. Finally, unlike the mean or median, the mode lacks many mathematical properties, making it less useful in further statistical analysis or inferential statistics.

> *Example* 4.4.1: **Mode of an Array of Numbers**
>
> Consider the following set of numbers: $2, 3, 3, 5, 7, 7, 7, 9, 11$. To find the mode, count the frequency of each number:
>
> - The number 2 appears once.
> - The number 3 appears twice.
> - The number 5 appears once.
> - The number 7 appears three times.
> - The number 9 appears once.
> - The number 11 appears once.
>
> Since the number 7 appears more frequently than any other number in the dataset, the mode of this set of numbers is 7.

4.5 Mid-Range

In statistics, the **mid-range** (sometimes called *mid-extreme*), often indicated with M, is a measure of central tendency of a sample defined as the arithmetic mean of the

4.5 Mid-Range

maximum and minimum values of the data (indicated here as usual with x).

$$M = \frac{\max x + \min x}{2} \quad (4.5)$$

The mid-range is a concept closely related to the **range** (which is defined as the difference between the maximum and minimum values in a dataset, indicating the spread of the data and defined in Sect. 5.3). These two measures work hand in hand, as knowing both the mid-range and the range allows one to deduce the highest and lowest values in the sample.

Definition 4.5.1: **Mid-Range**

The **mid-range** is defined as the arithmetic mean of the maximum and minimum values of the data

$$M = \frac{\max x + \min x}{2} \quad (4.6)$$

The mid-range is rarely utilised in practical statistical analyses. This is due to its limitations, since it disregards all values between the extremes and is highly sensitive to outliers.[1] For a broad range of distributions, it is considered one of the least effective and least resilient statistics. Nevertheless, the mid-range has its uses: It is, for example, the most efficient estimator for identifying the centre of a uniform distribution.

Example 4.5.1: **Mid-Range of an Array of Numbers**

Consider the dataset: 4, 8, 15, 16, 23, 42. To evaluate the mid-range we need to follow the next steps:

1. Identify the minimum value: 4.
2. Identify the maximum value: 42.
3. Calculate the mid-range: $(4 + 42)/2 = 23$.

The mid-range of this dataset is 23. Recall that while the mid-range is easy to calculate, it has several limitations. The mid-range only considers the extreme values of the dataset, making it highly sensitive to outliers. For example, if the dataset were 4, 8, 15, 16, 23, 100, the mid-range would increase significantly to 52, although most data points are much lower. Furthermore, the mid-range does not account for the distribution of the rest of the data. Whether the other numbers are clustered near the minimum or the maximum or spread out evenly, the mid-range remains the same. To add another limitation, recall that for skewed distributions or datasets with outliers, the mid-range can

---

[1] To be precise the mid-range is evaluated **only** with outliers if they are present, the maximum and the minimum.

be a misleading measure of central tendency, not accurately reflecting the dataset's typical values.

4.6 When to Use Mean, Median or Mode

The choice of which measure to use is summarised in the next info box.

> *Tip* 4.6.1: **When to Use Mean, Median, or Mode**
>
> Mean, median, and mode are measures of central tendency that summarise key aspects of a dataset. Choosing the appropriate measure depends on the nature of the data and the specific insight you are seeking to gain.
>
> - **Mean**: The **mean**, or average, is best used with numerical data where the values are evenly distributed without extreme outliers. It provides a useful overall measure of the central tendency when the data are symmetric. However, it can be misleading if the data contain outliers, as these can skew the mean.
> - **Median**: The **median** is the middle value of a dataset when it is ordered from lowest to highest and is less affected by outliers and skewed data. It is particularly useful when dealing with skewed distributions or ordinal data, where the mean may not accurately represent the central tendency. The median gives a better indication of the typical value in such cases.
> - **Mode**: The **mode** is the most frequently occurring value in a dataset. It is the only measure of central tendency that can be used with **categorical** data. The mode is especially useful for categorical data to determine the most common category. It can also be helpful to understand the distribution of data in addition to the mean or median.
>
> Choosing between mean, median, and mode depends on the data's distribution and the presence of outliers. For symmetric distributions without outliers, the mean is often preferred. For skewed distributions or when outliers are present, the median provides a more accurate reflection of the central tendency of the dataset. The mode is most useful for categorical data and for identifying the most common value in a dataset.
>
> A good example is the assessment of a **typical** salary in Switzerland. Since there are many people in Switzerland who have a very high salary (the CEO of IKEA, for example), using the mean would give a much skewed impression on what is a typical salary in Switzerland. So, the median is used instead, since it is less sensible to few outliers.

Chapter 5
Measures of Dispersion

5.1 Variance

When analysing data, one important aspect is understanding how **spread** they are around the mean or median. For example, consider the case of grades in a school. Imagine that the grades go from 0 to 100 and that the average (mean) is 78. There is a huge difference if the grades in the school go from 10 to 100 or if they go from 70 to 90. The latter case indicates that students perform generally well in the school, while the former case points to problems with students or teaching.

There are several ways of measuring how **spread** values are. The most important is the **variance** that is defined by the following formula:

$$\text{Var}(x) = \sigma^2(x) = \frac{1}{N} \sum_{i=1}^{N} (x_i - \mu)^2 \tag{5.1}$$

where σ^2 indicates the variance, x_i represents each data point in the dataset, μ is the mean of the dataset, and N is the number of data points in the dataset. The variance measures how far each number in the set is from the mean. A high variance indicates that the data points are spread out widely around the mean, while a low variance indicates that the data points are clustered closely around the mean.

Definition 5.1.1: **Variance**

The variance is defined by

$$\text{Var}(x) = \sigma^2(x) = \frac{1}{N} \sum_{i=1}^{N} (x_i - \mu)^2 \tag{5.2}$$

where σ^2 is used to indicate the variance, x_i represents each data point in the dataset, μ is the mean of the dataset, and N is the number of data points in the dataset.

It is important to differentiate between the variance of the population (indicated with σ^2) and the variance of the sample (indicated with S^2 or s^2). The formula given above calculates the population variance, where N is the size of the population. When calculating the sample variance (which we indicate here with S^2), the denominator should be $N - 1$ instead of N, to correct for the bias in estimating a population parameter from a sample (this is not completely correct but provides a good intuitive understanding, and a more precise explanation is given in the info box below).

$$S^2 = \frac{1}{N-1} \sum_{i=1}^{N} (x_i - \mu)^2 \tag{5.3}$$

Tip 5.1.1: ★ N or $N-1$ in the Variance Formula?

To answer this question, consider the random sample $x_1, x_2, \ldots, x_N$ from a population with mean μ and variance $\sigma^2 < \infty$. Then let us calculate the expectation value of the variance given in Eq. (5.3) (remember the expected value is the value you would expect on average if you would sample infinitely many times the N values x_i).

$$\begin{aligned}
\mathbb{E}(S^2) &= \mathbb{E}\left(\frac{1}{N-1} \left[\sum_{i=1}^{N} (x_i - \bar{x})^2 \right] \right) \\
&= \mathbb{E}\left(\frac{1}{N-1} \left[\sum_{i=1}^{N} (x_i^2 + \bar{x}^2 - 2x_i \bar{x}) \right] \right) \\
&= \mathbb{E}\left(\frac{1}{N-1} \left[\sum_{i=1}^{N} x_i^2 - N\bar{x}^2 \right] \right) \\
&= \{\text{Note that } \mathbb{E}(x_i^2) \text{ does not depend on } i\} \\
&= \frac{1}{N-1} (N\mathbb{E}(x_1^2) - N\mathbb{E}(\bar{x}^2)) \\
&= \frac{1}{N-1} \left(N(\sigma^2 + \mu^2) - N \left(\frac{\sigma^2}{N} + \mu^2 \right) \right) \\
&= \{\text{Using Eq. (5.2)}\} \\
&= \sigma^2
\end{aligned} \tag{5.4}$$

where we have used the results (which we will not prove here)

$$\mathbb{E}(\bar{x}) = \mu \tag{5.5}$$

and
$$\text{Var}(\bar{x}) = \frac{\sigma^2}{N} \tag{5.6}$$

In fact note that

$$\begin{aligned}
\mathbb{E}(\bar{x}^2) &= \text{Var}(\bar{x}) + \mathbb{E}(\bar{x})^2 \\
&= \{\text{Using Eqs. (5.5) and (5.6)}\} \\
&= \frac{\sigma^2}{N} + \mu^2
\end{aligned} \tag{5.7}$$

and

$$\begin{aligned}
\mathbb{E}(x_i^2) &= \text{Var}(x_i) + \mathbb{E}(x_i)^2 \\
&= \sigma^2 + \mu^2
\end{aligned} \tag{5.8}$$

So, from this calculation, it is clear that $\mathbb{E}(S^2)$ will give you the variance of the **population**. This is what is called an *unbiased* estimator of population variance. If you consider Eq. (5.2) and evaluate its value over many measurements and calculate the expected value, you would end up with a value that will not estimate the correct variance (the one from the population), but that would be slightly off. This is the real reason why $N-1$ is needed in Eq. (5.3). In other words you would have

$$\mathbb{E}\left(\frac{1}{N}\sum_{i=1}^{N}(x_i - \mu)^2\right) = \frac{N-1}{N}\sigma^2 \tag{5.9}$$

which is **not** the variance of the population.

5.2 Standard Deviation

If you understand the variance, you will understand the standard deviation (typically indicated with σ). In fact we have the relationship

$$\sigma(X) = \sqrt{\text{Var}(X)} \tag{5.10}$$

or in other words

$$\sigma(X)^2 = \text{Var}(X) \tag{5.11}$$

It is important to underline that the standard deviation is measured in the same units as the data and thus is easier to interpret. It indicates, on average, how far each data point is from the mean.

Definition 5.2.1: **Standard Deviation**

The standard deviation σ is defined by

$$\sigma(X)^2 = \text{Var}(X) \tag{5.12}$$

The standard deviation is measured in the same units as the data.

It also has a special meaning in the case of a normal distribution (more on that in Sect. 8.3).

Example 5.2.1: **Variance and Standard Deviation**

Consider the dataset: 3, 7, 7, 19. First, we need to calculate the mean (μ) of the dataset:

$$\mu = \frac{3 + 7 + 7 + 19}{4} = 9$$

Next, we calculate the variance (σ^2) using the formula

$$\sigma^2 = \frac{1}{N} \sum_{i=1}^{N} (x_i - \mu)^2$$

For our dataset we have

$$\sigma^2 = \frac{(3-9)^2 + (7-9)^2 + (7-9)^2 + (19-9)^2}{4} = \frac{36 + 4 + 4 + 100}{4} = 36$$

Finally, calculate the standard deviation (σ) as the square root of the variance

$$\sigma = \sqrt{\sigma^2} = \sqrt{36} = 6$$

For the given dataset 3, 7, 7, 19, the variance is 36, and the standard deviation is 6.

If we calculate the sample variance S^2, we would get

$$S^2 = \frac{(3-9)^2 + (7-9)^2 + (7-9)^2 + (19-9)^2}{3} = \frac{36 + 4 + 4 + 100}{3} = 48$$

which, as you may notice, is much higher than 36 but that is known to be an unbiased estimate of the population variance (which is not 36, as this too has been estimated from the population data) that is unknown.

Tip 5.2.1: **Sample Variance** S^2

To make the story short, when you want to estimate the variance of a population from a sample, always use the S^2 formula given by

$$S^2 = \frac{1}{N-1} \sum_{i=1}^{N} (x_i - \mu)^2$$

5.3 Range

The **range** in statistics is a measure of dispersion or variability that indicates the difference between the highest and lowest values in a dataset. It is calculated simply as

range = Maximum value − Minimum value

The range gives a quick sense of the spread of a dataset and can be used to understand the extent of variability among observed values. Although it is straightforward to calculate and easy to understand, the range is highly sensitive to extreme values (outliers) because it only considers the two extreme values in the dataset (maximum and minimum).

Definition 5.3.1: **Range**

The **range** in statistics measures the difference between the highest and lowest values in a dataset. It is defined simply as

range = Maximum value − Minimum value

Despite its simplicity, the range has several limitations. First, it does not provide information about the distribution of values between the two extremes. Second, as mentioned already, it is sensitive to extreme values (outliers), which can significantly skew the range, making it less representative of the dataset as a whole. Furthermore, it does not account for the size of the dataset, making comparisons between datasets of different sizes potentially misleading.

The range is often used in preliminary data analysis to get a basic understanding of the spread of the data. It can be particularly useful in contexts where the maximum and minimum values are of specific interest, such as quality control processes. However, for a more detailed analysis of the dispersion of data, other measures such as the interquartile range, variance, or standard deviation are typically preferred due to their ability to provide more information about the overall distribution of the data.

5.4 Dangers of Relying on Single Statistics

We have looked at measures of central tendencies and dispersions (e.g. mean and variance). It is tempting to simply give those two numbers and think that they describe your data sufficiently well. This is quite dangerous. To convince you, I would like you to check Fig. 5.1. Each panel shows a different dataset of tuples (x_i, y_i), and at the top the mean of x and y (indicated with μ_x and μ_y, respectively) and the standard deviation of x and y (indicated with σ_x and σ_y, respectively) are reported. You should notice that the statistical properties are the same for all the datasets in the different panels, even when visually they are dramatically different.

The dataset (called *Datasaurus*, and I am sure you can guess why) is described in [12] and was created by Justin Matejka and George Fitzmaurice[1] to show how wildly different (at least visually) datasets may have the same (specific) statistical properties. In their paper, they presented a technique for creating visually dissimilar datasets.

> *Warning* 5.4.1: **Dangers of Relying on Single Statistics**
>
> It is dangerous to use single statistics assuming that they are enough to describe the data. This is why it is important to study the distribution of data, visualise it, and understand the measures of position (see Chap. 6).
>
> For example, data can be bimodal (having two peaks) (we discuss modality in Sect. 9.4), skewed (we discuss skewness in Sect. 9.2), or even uniformly distributed yet still have the same mean and variance. Without considering the shape of the distribution, decisions made on the basis of single statistics can be fundamentally flawed.
>
> Single statistics are fundamentals, but they should be regarded as part of a broader data analysis process that includes a visual and statistical examination of the entire data distribution and its characteristics.

---

[1] See https://www.openintro.org/data/index.php?data=datasaurus for the dataset.

5.4 Dangers of Relying on Single Statistics

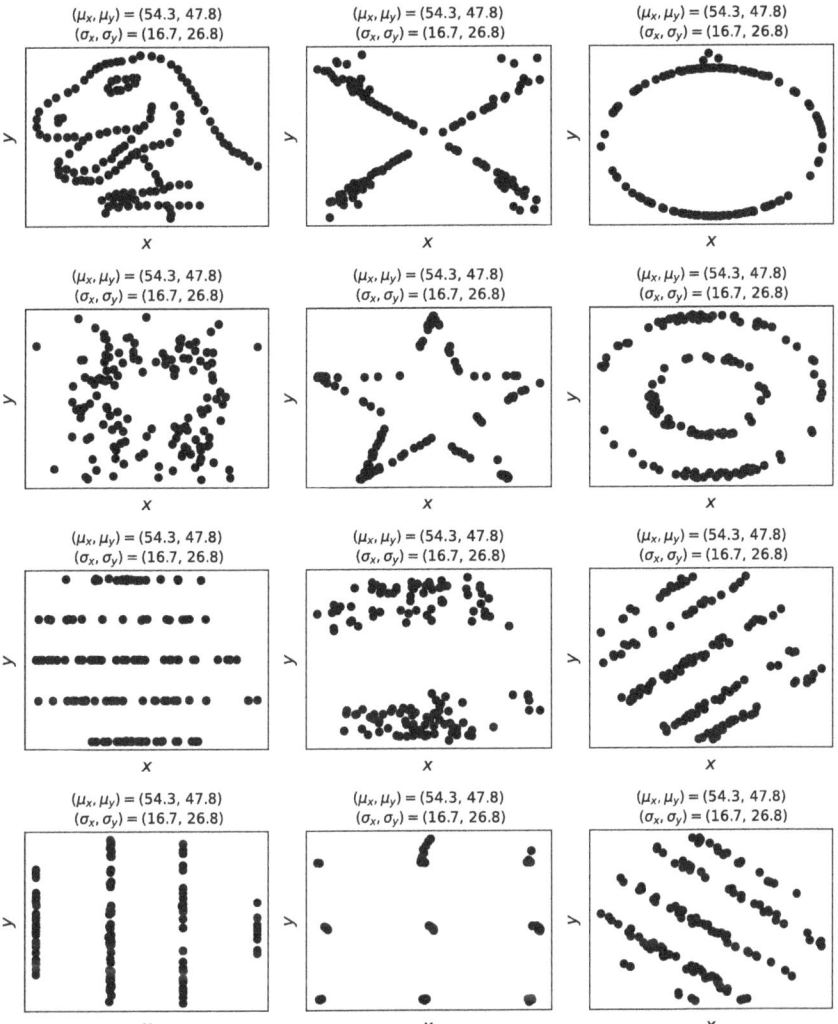

Fig. 5.1: Each panel shows a different dataset of tuples (x_i, y_i), and at the top the mean of x and y (indicated with μ_x and μ_y, respectively) and the standard deviation of x and y (indicated with σ_x and σ_y, respectively) are reported. The data is described in [12] and were created by Justin Matejka and George Fitzmaurice to show how wildly different datasets may have the same (specific) statistical properties

Chapter 6
Measures of Position

6.1 Introduction

Measures of position (sometimes known as location measures) are used to describe the position or rank of individual values within a dataset relative to the entire distribution. These measures provide insight into the structure of the dataset, helping to identify where specific data points lie in the distribution.

Measures of position are important in statistics because they provide insight into the distribution and structure of a dataset. By evaluating where specific data points lie relative to the entire distribution, they help in understand how data are spread around its values. For instance, percentiles, quartiles, and deciles divide the dataset into parts, allowing us to see where a particular value is located. This is especially useful for identifying outliers, trends, or shifts in the data. Another key reason for using measures of position is to facilitate comparison between datasets or observations within a dataset. For example, when comparing test scores, knowing that a score is in the 90th percentile indicates that it is higher than 90% of the other scores. This provides a lot more information about performance than simply looking at the single score.

Furthermore, measures of position offer a way to summarise data without being overly influenced by extreme values. Unlike the mean, which can be skewed by outliers (see Chap. 4), measures like the median or interquartile range (IQR) are more robust. This makes them useful for analysing skewed data or distributions with outliers, where central tendency measures alone might be misleading. In general measure of position offers a deeper and more nuanced view of data beyond simple averages.

6.2 Percentiles

Percentiles are numerical values that divide a dataset into 100 equal parts, with each part representing 1% of the distribution. The qth percentile is the value below which q% of the data falls. More precisely the qth percentile is a value **below** which a given percentage q of values falls (**exclusive** definition) or a value at **or** below which a q percentage falls (**inclusive** definition). Note that using an **exclusive** or **inclusive** definition will give you different results. Note also that percentiles are usually expressed in the same units of the data, not in percent. For example, if you have a dataset of lengths of certain mechanical parts measured in centimetres, the 10th percentile could be, say, 23 cm.

There are many algorithms to calculate percentiles, and different software tools (like Microsoft Excel, Python NumPy, or R) do it in different ways. Hyndman and Fan identified nine different ways in which common software packages calculate percentiles [13]. You should always check in the documentation how a specific software tool calculates percentiles and declare the method in your own work. Generally speaking, algorithms either return the value of an element that is in the dataset (nearest-rank method described in Sect. 6.2.1) or interpolate between existing values of elements in the dataset. Furthermore, methods are either **exclusive** or **inclusive** (as briefly discussed at the beginning of the section).

To summarise the discussion, when you calculate the qth percentile, two things can happen (depending also on the definition you use, exclusive or inclusive and on the algorithm you use). Your value may lie between two existing data points (panel (a) in Fig. 6.1) or match exactly one data point (panel (b) in Fig. 6.1). If the qth percentile does not match any precise value of the array v, then some interpolation method is necessary since no value in the dataset matches the position of the qth percentile.

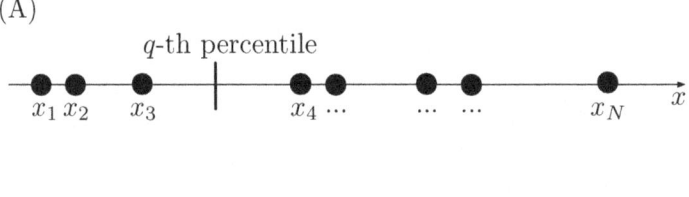

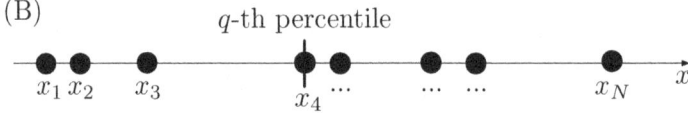

Fig. 6.1: When you calculate the qth percentile, two things can happen. Your value lies between two existing data points (case (A)) or be exactly one data point (case (B))

6.2.1 Nearest-Rank Method

Probably the simplest way to calculate percentiles is to use the **nearest-rank** method. To get the kth percentile in a dataset with N data points, you follow the steps below:

1. Calculate the ordinal rank r

$$r = \left\lceil \frac{k}{100} N \right\rceil \quad (6.1)$$

where the symbol $\lceil x \rceil$ is the ceiling of x, that is the smallest integer that is not smaller than x.

2. Choose the value in the dataset at the rth position (x_r). This is your kth percentile.

> *Example* 6.2.1: **Percentiles with the Nearest-Rank Method**
>
> Let us make an example. Consider the following array ($N = 13$):
>
> $$v = \{1, 3, 4, 7, 10, 12, 15, 20, 22, 25, 30, 32, 37\}$$
>
> Suppose that we want to know the 10th percentile by using the exclusive definition ($k = 10$). Following the algorithm described in the text, we have
>
> $$r = \left\lceil \frac{10}{100} 13 \right\rceil = \lceil 1.3 \rceil = 2 \quad (6.2)$$
>
> thus our 10th percentile is $v_2 = 3$. The symbol $\lceil x \rceil$ is the ceiling of x, that is the smallest integer that is not smaller than x.

6.2.2 Linear Interpolation Between Ranks

This explanation is slightly more convoluted, so take your time reading this section. In the end, I will try to summarise it in more intuitive terms. Consider having a sorted array

$$v = \{v_1, v_2, \ldots, v_N\} \quad (6.3)$$

with $v_{i+1} \geq v_i \; \forall i = 1, \ldots, N-1$. We want a linear interpolation function

$$v(x) = v_{\lfloor x \rfloor} + (x \bmod 1)(v_{\lfloor x \rfloor + 1} - v_{\lfloor x \rfloor}) \quad (6.4)$$

where $x \in [1, N]$, and $\lfloor x \rfloor$ is the integer part of x and ($x \bmod 1$) the fractional part of x. For example $\lfloor 2.3 \rfloor = 2$, and ($2.3 \bmod 1$) = 0.3. Furthermore, note that for $x = i \in \mathbb{N}$, we have $v(i) = v_i$. So in a sense x is the continuous version of the index i. Sometime x is called the *virtual* index.

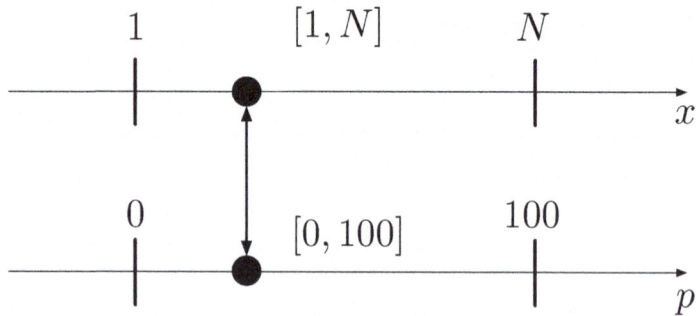

Fig. 6.2: A visual representation of the mapping between the virtual index x and p

---

Tip 6.2.1: **Virtual Index Notation**

The virtual index x is sometimes indicated with $i + g$ where

$$i = \lfloor x \rfloor \tag{6.5}$$

and

$$g = (x \bmod 1) \tag{6.6}$$

In other words, i is the integer part of x and g the fractional part. For example if $x = 2.3$, we would have $i = 2$ and $g = 0.3$.

Now suppose that we are looking at the kth percentile (e.g. the 10th percentile). We can define $p = k/100$. Then we need to map the virtual index x to the range $[0, 100]$. We can do that with a linear transformation (see for an intuitive visualisation Fig. 6.2). This can be done with a linear relationship

$$x = (N + c_1)p + c_2 \tag{6.7}$$

Imposing that the midpoint of the x range $[1, N]$ that is located at the index $(N+1)/2$ occurs at $p = 0.5$, we obtain

$$\frac{N+1}{2} = \frac{N+c_1}{2} + c_2 \tag{6.8}$$

and thus $2c_2 + c_1 = 1$. So we have just one constant that we can choose. In other words we can write

$$x = (N + 1 - 2c_2)p + c_2 \tag{6.9}$$

Now the different software packages use different values for c_2.

6.2 Percentiles

Warning 6.2.1: **Percentiles in numpy**

In Python you have the numpy.percentile() function that will evaluate the percentiles. To use it, you need to specify the method for interpolating between elements of the input array or in other words how to calculate the virtual index x. There are many methods available as linear, inverted_cdf, hazen, and weibull between others.

The different methods evaluate the virtual index in different ways. In the function numpy.percentile() the virtual index x is dependent on two parameters, α and β, and is given by the formula

$$x = i + g = \frac{q}{100}(N - \alpha - \beta + 1) + \alpha \tag{6.10}$$

where q is the percent rank (e.g. if we are looking at the 10th percentile, then $q = 10$). You can specify different methods when using the Python function, and each uses different values for α and β. Note that for many it is not true that the midpoint of the range $[1, N]$ lies at $p = 50$, so be warned. The methods are those described in the paper by Hyndman and Fan [14].

For example choosing weibull as parameter, the value chosen for both α and β is zero, thus giving the relationship

$$x = i + g = \frac{q}{100}(N + 1) \tag{6.11}$$

while choosing linear as method, the value chosen for α and β is 1, thus giving the relationship

$$x = i + g = \frac{q}{100}(N - 1) + 1 \tag{6.12}$$

You must be aware of the method you are using to calculate percentiles. Percentiles are not uniquely defined, and care must be taken in their evaluation. Keep in mind that while using any statistical function from a software package (or programming language) is always important to check the documentation, and check how percentiles are evaluated.

Typical choices for c_2 are 1/2 (used in Matlab), 1 (equivalent to the weibull method to calculate percentiles in the Python numpy library) or 0, which is the value recommended by the National Institute of Standards and Technology, NIST (see https://www.itl.nist.gov/div898/handbook/prc/section2/prc262.htm for more information).

To summarise, the simplest approach to evaluate the percentile is composed of two steps: (i) evaluate the *virtual* index x and (ii) evaluate the percentile itself. As an example let us consider the virtual index given by the formula (the one for $c_2 = 0$)

$$i + g = \frac{(N + 1)q}{100} \tag{6.13}$$

where N is the total number of elements of the array v. Consider the array $v = (1, 2, 3, 4, 5)$. We have

$$i + g = \frac{6 \times 25}{100} = 1.5 \qquad (6.14)$$

so $i = 1$ and $g = 0.5$. The virtual index lies between the first ($v_1 = 1$ in our example) and second ($v_2 = 2$ in our example) elements of the array v. Since we do not have any information about how data are distributed between 1 and 2, we need to make some assumptions. We can do a linear interpolation, and thus we would find (see again Eq. (6.4))

$$25\text{th percentile} = v_1 + g(v_2 - v_1) = 1 + 0.5(2 - 1) = 1.5 \qquad (6.15)$$

So the 25th percentile of $v = (1, 2, 3, 4, 5)$ is 1.5. Or said in other words, 25% of the data lies below 1.5. As you have seen in the example above, the qth percentile often falls between elements of your array. In this case, you need to make some assumptions on how the data behave between your elements (e.g. by choosing c_2). There are many variations on how to do that, and a nice overview, as mentioned already, can be found in [14]. A linear interpolation may be fine if the data points are close to each other, but what about if we have $v = (1, 100, 101, 102, 103, 104)$? We still have $i + g = 1.5$, but can we really be sure that between 1 and 100 is data well approximated by a linear interpolation? Much can happen in such a wide range. There is never always a clear receipt to decide on how to interpolate, and some information on the data itself and some educated guess are necessary to choose the right method.

> *Example* 6.2.2: **Percentiles**
>
> Given a dataset
> $$\{22, 25, 28, 31, 34, 37, 40, 43\}$$
> we want to calculate the 25th, 50th (median), and 75th percentiles. The dataset is already sorted in ascending order. The formula to find the virtual index for the kth percentile (P_n) is
>
> $$\frac{(N + 1) \times k}{100}$$
>
> where N is the total number of data points, and k is the percentile number. Consider now the task of calculating the 25th percentile. The virtual index is now
>
> $$\frac{(8 + 1) \times 25}{100} = 2.25$$

6.2 Percentiles

That means that the 25th percentile lies between the second and third elements of the dataset. Since P_{25} is not an integer, we must interpolate between the second and third data points (25 and 28). Using Eq. (6.4) we have

$$\text{25th percentile} = 25 + 0.25 \times (28 - 25) = 25.75$$

To clarify the concept again, let us calculate the 50th percentile (nothing else than the media). For the 50th percentile the virtual index x is given by

$$\frac{(8+1) \times 50}{100} = 4.5$$

Interpolating between the fourth and fifth data points (31 and 34):

$$\text{50th percentile} = 31 + 0.5 \times (34 - 31) = 32.5$$

If you remember how to calculate the median, you will notice that you will get the same result. In fact, this dataset has eight numbers, so we find the median by averaging the fourth and fifth numbers (see Sect. 4.3). The fourth number is 31, and the fifth number is 34. To calculate the median, we average these two numbers

$$\text{Median} = \frac{31 + 34}{2} = 32.5 \qquad (6.16)$$

which is exactly the same result for the 50th percentile.

Finally, let us calculate the 75th percentile. The virtual index x is now

$$\frac{(8+1) \times 75}{100} = 6.75$$

interpolating between the sixth and seventh data points (37 and 40)

$$\text{75th percentile} = 37 + 0.75 \times (40 - 37) = 39.25$$

For our dataset, the 25th percentile is 25.75 years, the 50th percentile (median) is 32.5 years, and the 75th percentile is 39.25 years. These calculations indicate that 25% of the data lies below than 25.75, 50% lies below than 32.5, and 75% lies below than 39.25.

Percentiles are useful for understanding the distribution of data across various levels. For example, the 50th percentile, also known as the median, divides the dataset into two equal halves. Percentiles are commonly used in standardised testing to compare an individual's performance against a broader population.

> *Definition 6.2.1*: **Percentiles**
>
> Percentiles are numerical values that divide a dataset into 100 equal parts, with each part representing 1% of the distribution.
> The kth percentile is a value below which a given percentage k of scores falls (**exclusive** definition) or a score at **or** below which a k percentage falls (**inclusive** definition).

6.3 Quartiles

Let us now turn our attention to quartiles. Quartiles are statistical measures that divide a dataset into four equal parts, each representing a quarter of the distributed sampled data. They are used to describe the spread and centre of the data, much like the median divides the data into two halves. The quartiles are denoted by $Q1$ (the first quartile), $Q2$ (the second quartile, also the median), and $Q3$ (the third quartile).

> *Definition 6.3.1*: **Quartiles**
>
> Quartiles are statistical measures that divide a dataset into four equal parts, each representing a quarter of the distributed sampled data. The quartiles are denoted by $Q1$ (the first quartile), $Q2$ (the second quartile, also the median), and $Q3$ (the third quartile).

The easiest way of determining quartiles is the following.

1. **Arrange the data** in ascending order.
2. **Find the median** ($Q2$): This is the middle value of the dataset. If there is an even number of observations, $Q2$ is the average of the two middle numbers (check again Sect. 4.3 for more information). This divides the dataset into two halves. This is also the 50th percentile.
3. **Find the first quartile** ($Q1$): This is the median of the lower half of the dataset (excluding $Q2$ if the number of observations is odd). $Q1$ represents the value below which 25% of the data falls. This is also the 25th percentile.
4. **Find the third quartile** ($Q3$): This is the median of the upper half of the dataset (excluding $Q2$ if the number of observations is odd). $Q3$ represents the value below which 75% of the data falls. This is also the 75th percentile.

> *Example 6.3.1*: **Quartiles**
>
> Consider the dataset: 2, 4, 4, 5, 7, 9, 11, 12, 14, 15, 17, 19. The data are already arranged in ascending order. We need to first find the median (Q2). Since

there are 12 numbers, the median is the average of the sixth and seventh numbers: $(9 + 11)/2 = 10$. Then we can find the first quartile (Q1). The first quartile is the median of the lower half of the dataset. The lower half (excluding the median for even-numbered datasets) is 2, 4, 4, 5, 7, 9. Thus, $Q1$ is the average of the third and fourth numbers: $(4 + 5)/2 = 4.5$. Then we need to find the third quartile (Q3). This is the median of the upper half of the dataset. The upper half is 11, 12, 14, 15, 17, 19. Thus, $Q3$ is the average of the third and fourth numbers: $(14 + 15)/2 = 14.5$.

For the given dataset, the quartiles are as follows:

- First quartile (Q1): 4.5
- Second quartile (Q2) or median: 10
- Third quartile (Q3): 14.5

These quartiles help in understanding the distribution of the data, indicating that 25% of the data is below 4.5, 50% is below 10, and 75% is below 14.5.

Quartiles are fundamental in descriptive statistics for understanding the distribution of data. They help in identifying the spread of the data by highlighting the range of the middle 50% of the values (between $Q1$ and $Q3$), known as the IQR (see Sect. 6.4). Furthermore, quartiles are used in the construction of boxplots, a type of graph that displays the distribution of data based on a five-number summary (*minimum*, $Q1$, *median*, $Q3$, and *maximum*) (more on that in Chap. 10). Finally, they provide insights into the symmetry and skewness[1] of the data distribution. For example, if $Q2$ is closer to $Q1$ than to $Q3$, the distribution is skewed to the right (we discuss skewness in Sect. 9.2).

Warning 6.3.1: **Calculations of Quartiles**

Note that the method described in this section is the simplest to calculate the quartiles. In theory, you can use percentiles to calculate quartiles. In any case, remember to specify in your studies and publications how you calculated measures of positions as quartiles.

6.4 Interquartile Range

Now we can turn our attention to the concept of the IQR, namely a measure of statistical dispersion, or variability, which indicates the spread of the middle 50% of data points in a dataset. Unlike range, which considers the difference between the maximum and minimum values, the IQR focuses on the central portion of the

---

[1] Skewness is a measure of how asymmetric a distribution of values is.

dataset, thereby providing a better sense of the overall variability of the data while being less sensitive to outliers.

IQR is calculated by subtracting the first quartile ($Q1$) from the third quartile ($Q3$):
$$IQR = Q3 - Q1$$

Recall that $Q1$ is the value below which 25% of the data falls, and $Q3$ is the value below which 75% of the data falls. The steps to calculate the IQR are as follows:

1. Arrange the data in ascending order.
2. Calculate $Q1$.
3. Calculate $Q3$.
4. Subtract $Q1$ from $Q3$ to find the IQR.

IQR is a robust measure of variability that is especially useful for identifying and summarising the spread of the middle half of a dataset. By focusing on the central 50% of the data, the IQR is less affected by extreme outliers or non-symmetric distributions of the data, making it a preferred choice over the range in many situations.

> *Definition* 6.4.1: **Interquartile Range**
>
> The IQR is calculated by subtracting the first quartile ($Q1$) from the third quartile ($Q3$).
> $$IQR = Q3 - Q1$$
> It is a measure of statistical dispersion, or variability, that indicates the spread of the middle 50% of data points in a dataset.

Moreover, IQR is commonly used in the construction of boxplots, where it visually represents the spread of the central data points and helps in the identification of outliers, which are typically defined as any data point that falls more than $1.5 \times IQR$ above $Q3$ or below $Q1$ (we discuss boxplots in Chap. 10).

6.5 Deciles

Sometimes you will encounter deciles, which are similar to percentiles, but they divide the dataset into ten equal parts, with each part representing 10% of the distribution. The nth decile is the point below which $n \times 10\%$ of the data lies.

Deciles offer a more detailed view of the distribution than quartiles but are less granular than percentiles.

> *Definition* 6.5.1: **Deciles**
>
> Deciles divide the dataset into ten equal parts, with each part representing 10% of the distribution. The nth decile is the point below which $n \times 10\%$ of the data lies.

6.6 Quantiles

Quantiles are numerical values that divide a dataset into parts, each having equal probabilities. In other words, a q-quantile is a value x at or below which a fraction q of the data lies and a fraction $1-q$ above x lies. You may wonder what the difference is between percentiles and quantiles. Percentiles are typically measured in the same unit of the data, while quantiles are given as a percentage.

In general, quartiles, percentiles, and deciles are all types of quantiles. Quartiles divide the distribution into four equal sections, percentiles into 100 equal sections, and deciles into ten equal sections.

Chapter 7
Outliers

7.1 Introduction

If you use statistics in your projects (and since you are reading this, you probably will), you will have to deal with **outliers**. But what are outliers? There are many definitions, but intuitively, outliers are observations in a dataset that are **significantly** different from the rest. What does "significantly" really mean? Here are a few possible definitions that you will find in the literature and in textbooks. Ultimately, how you define outliers depends on the problem you are trying to solve, on your data, and on your analysis.

7.2 Interquartile Range (IQR) Method

As usual, we will imagine to have a dataset composed of N values $\{x_i\}_{i=1}^{N}$. The **IQR method** defines an outlier as a value x_i below $Q1 - 1.5 \times \text{IQR}$ or above $Q3 + 1.5 \times \text{IQR}$, where $Q1$, $Q3$, and IQR are the first and third quartiles and the interquartile range, respectively. This method is often used because it is less influenced by extreme values. This is also the method that is typically used to mark points as outliers in boxplots. In fact, in a boxplot, typically (but not always) data points outside the whiskers (see Chap. 10), which typically extend to $1.5 \times \text{IQR}$ from the first and third quartiles, are considered outliers.

> *Definition* 7.2.1: **Outliers Defined with the IQR**
>
> The **IQR method** defines an outlier as any data point with a value below $Q1 - 1.5 \times IQR$ or above $Q3 + 1.5 \times IQR$, where $Q1$, $Q3$, and IQR are the first and third quartiles and the interquartile range, respectively.
> The isolated points you see often in boxplots are typically defined with this approach (see Chap. 10).

7.3 Domain-Specific Criteria

Some fields define outliers according to specific knowledge or criteria relevant to the subject matter, recognising that what constitutes an outlier can vary by context.

> *Example* 7.3.1: **Domain-Specific Criteria for Outliers**
>
> Here are some examples of domain-specific criteria to define outliers.
> **Industry Standards**: Outliers could be defined as companies whose annual revenue deviates significantly (here it needs to be decided what significantly means) from the average revenue for start-ups in the same industry and of similar size.
> **Funding Rounds**: Outliers could be identified based on the amount of funding they have raised compared to other start-ups in the same stage of development. For example, a start-up that has raised ten times more funding than its peers might be considered an outlier. For example, consider what is called a *unicorn* between start-ups (a unicorn refers to a start-up with a valuation exceeding 1 billion USD), which is a very good example of how outliers are defined according to some domain-specific criteria.
> **Geographical Considerations**: Outliers could also be defined based on geographical factors such as location-specific market conditions or regulatory environments. A start-up operating in a region with vastly different economic conditions or consumer behaviours might be treated as an outlier.

7.4 z-Score Method

A data point with a z-score (a measure of how many standard deviations an element is from the mean) beyond a certain threshold, such as 3 or -3, can be considered an outlier. This method quantifies the distance from the mean in standard deviations.

7.4 z-Score Method

Definition 7.4.1: **Outliers Defined with the z-Score**

A data point with a z-score (a measure of how many standard deviations an element is from the mean) beyond a certain threshold, such as 3 or −3 (the choice of this value is somewhat subjective and depends on the problem) is considered an outlier.

Here, we will give you a short definition of what the z-score is. The **z-score**, also known as a standard score, quantifies the number of standard deviations a data point is from the mean of a dataset. It is a measure of how unusual or typical a data point is compared to the average of the dataset. Imagine that we have a dataset and a data point from it x. The formula to calculate a z-score is the following:

$$z = \frac{(x - \mu)}{\sigma}$$

where x is the value of the data point, μ is the mean of the dataset, and σ is the standard deviation of the dataset. The z-score is a critical tool in statistics for several reasons. First, it allows for comparison between data points from different datasets by normalising the data. Second, a z-score of 0 indicates that the value of the data point is identical to the mean value. Third, positive z-scores indicate values greater than the mean, while negative z-scores indicate values less than the mean. Furthermore, z-scores can identify outliers in a dataset. Often, data points with a z-score greater than +3 or less than −3 are considered outliers by many statisticians.

Example 7.4.1: **z-Score**

Suppose we have a dataset representing the test scores of a class: 82, 90, 76, 94, and 88. We want to calculate the z-score for a test score of 88. First, calculate the mean (μ) of the dataset:

$$\mu = \frac{82 + 90 + 76 + 94 + 88}{5} = 86$$

Next, calculate the standard deviation (σ). The standard deviation formula is

$$\sigma = \sqrt{\frac{\Sigma(X - \mu)^2}{N}}$$

where $N = 5$. For simplicity, we give only the result here:

$$\sigma \approx 6.633$$

Finally, we can calculate the z-score using the formula:

$$z = \frac{(X - \mu)}{\sigma}$$

Substituting the values

$$z = \frac{(88 - 86)}{6.633} \approx 0.302$$

The z-score for a test score of 88 is approximately 0.302. This means the score is 0.302 standard deviations above the mean score of the class. A z-score close to 0 indicates that the data point is close to the mean, reflecting that a score of 88 is relatively average in this dataset and is higher than μ.

Warning 7.4.1: z-Score for Not Normally Distributed Data

Note that if the data are not normally distributed, the use of the z-score should be used with care. Skewed or multimodal distributions (see Chap. 8) will make the interpretation of the score less obvious. If you are dealing with data that are asymmetric and skewed, be careful in interpreting the z-score.

7.5 Causes, Impact, and Treatment

Outliers can occur due to various reasons, such as data entry errors, measurement errors, natural variation, or the presence of anomalous observations in the population being studied. Human errors during data entry, such as typos or transcription errors, can lead to outliers. For example, entering a decimal point in the wrong place can result in a vastly different value. Furthermore, errors in measurement instruments or techniques can produce outliers. Variability in measurement devices, calibration issues, or environmental factors can contribute to measurements that are significantly different from the rest of the data. Particularly important is when outliers can arise due to sample variability, especially in smaller sample sizes. If the sample is not representative of the population or if there are unusual characteristics in the sampled individuals, outliers may find their way in your dataset. Sometimes, outliers represent genuine extreme values in the data (remember the COVID pandemic, for example?) These could be the result of rare events, unusual circumstances, or outliers that are truly indicative of important phenomena being studied. While less common, genuine extreme values can still influence statistical analyses and interpretations.

Always look at your data for outliers, as they can have a significant impact on statistical analyses, especially those that are sensitive to extreme values such as the mean and standard deviation. They can distort the measures of central tendency and dispersion, leading to biased (or unclear) results. Depending on the nature of the data and the research question, outliers can be dealt with in different ways. In some cases, outliers may be removed from the dataset if they are deemed to be due to errors or if they significantly affect the results (maybe if you are studying the income in Switzerland you can leave the IKEA CEO out of your study, he is probably not so representative of the average worker in Switzerland). Transforming the data using techniques such as logarithmic transformation can sometimes mitigate the effects of outliers but is a tricky thing to do and keep under control. A good approach is using robust statistical methods that are less affected by outliers, such as the median instead of the mean or non-parametric tests, which can be appropriate in certain situations. Regardless of the treatment chosen, it is important to **transparently report** any outliers and the rationale behind their treatment in research findings.

Chapter 8
Introduction to Distributions

8.1 A Small Warning

Distributions are at the core of statistics, but their understanding and study require more mathematics. To follow this and the following chapters you will need to have a good grasp of calculus (especially integration and derivation). I tried to always give also intuitive descriptions of concepts, but I cannot deny that this chapter is more mathematically heavy. Take your time to study it, since it is something every statistician should know.

8.2 Introduction to Probability Distributions

Imagine you are conducting an experiment, tossing a fair coin multiple times and recording the number of times it lands heads-up. Each time you toss the coin, the outcome (heads or tails) is random, but over many tosses, you start to notice patterns in the results. Some outcomes, like getting exactly half heads and half tails over a large number of tosses, are more common than others (like getting a long series of heads or tails one after the other). A **probability distribution** captures these patterns by describing the likelihood of each possible outcome of a random experiment. It tells you how probable it is to observe each outcome, given the rules of the experiment and any underlying randomness.

> *Definition* 8.2.1: **Probability Distribution (Intuitive Definition)**
>
> A **probability distribution** describes the likelihood of each possible outcome of a random experiment. It tells you how probable it is to observe each outcome, given the rules of the experiment and any underlying randomness.

In some experiments, the outcomes are distinct and countable, like rolling a six-sided die or counting the number of heads in a series of coin tosses. For these experiments, we use **discrete probability distributions** (see Sect. 8.2.1). Imagine you are rolling a fair six-sided die. The probability distribution for this experiment assigns a probability to each possible outcome: rolling a 1, rolling a 2, and so on up to rolling a 6. Each outcome has a probability of 1/6 because there are six equally likely outcomes (assuming a fair dice).

Other experiments involve outcomes that can take on any value within a range, like measuring the height of people or the time it takes for a chemical reaction to occur. For these experiments, we use **continuous probability distributions** (see Sect. 8.2.2). Imagine you are measuring the height of people in a population. The probability distribution for this experiment describes the likelihood of observing different heights. Some heights, like the average height, are more common, while extreme heights are less likely (few individuals reach a height of 2 m, whereas the majority fall within the 1.6–1.8 m range). The probability density function (PDF) (see Sect. 8.2.2) in this case describes this distribution by specifying how the probability is distributed over the range of possible values. For example, the PDF might show that heights around the average are more probable, while very tall or very short heights are less probable.

Probability distributions provide a way to understand and quantify uncertainty in random experiments. They allow us to make predictions about the likelihood of different outcomes.

8.2.1 Discrete Probability Distribution

A discrete probability distribution is applicable when the random variable can only take on a finite or countably[1] infinite number of distinct values. The **probability mass function** (PMF) is used to describe the probability distribution of a discrete random variable. For a discrete random variable X, the PMF is denoted by $p(x)$, where $p(x)$ represents the probability that X takes on the value x.

> *Definition* 8.2.2: **Probability Mass Function**
>
> The PMF is used to describe the probability distribution of a discrete random variable. For a discrete random variable X, the PMF is denoted by $p(x)$, where $p(x)$ represents the probability that X takes on the value x.

---

[1] A set of objects is countable if either it is finite or it is in one-to-one correspondence with $\mathbb{N}$. A wonderful explanation of this concept can be found in the book by Abbott *Understanding Analysis* [15].

8.2 Introduction to Probability Distributions

Example 8.2.1: **Probability Mass Function: Rolling a Six-Face Dice**

When rolling a fair six-sided dice, the possible outcomes are the numbers 1 through 6. Each outcome has an equal probability of 1/6, assuming the die is fair. The PMF for this experiment assigns a probability to each possible outcome. Let X be the random variable that represents the outcome of the die roll. The PMF of X is then given by

$$p(x) = \begin{cases} 1/6, & \text{if } x = 1, 2, 3, 4, 5, \text{ or } 6 \\ 0, & \text{otherwise} \end{cases}$$

This PMF indicates that each outcome x has a probability of 1/6, as there are six equally likely outcomes when rolling a fair six-sided dice.

8.2.2 Continuous Probability Distribution

A continuous probability distribution is applicable when the random variable can take on any value within a specified range. The **probability density function** (PDF) is used to describe the probability distribution of a continuous random variable. For a continuous random variable X, the PDF is denoted by $f(x)$, where $f(x)$ represents the probability density at the point x. The probability of X lying within a given interval is obtained by integrating the PDF over that interval.

Definition 8.2.3: **Probability Density Function**

The PDF is used to describe the probability distribution of a continuous random variable. For a continuous random variable X, the PDF is denoted by $f(x)$, where $f(x)$ represents the probability density at the point x.
The probability of X lying within a given interval is obtained by **integrating** the PDF over that interval.

Warning 8.2.1: **PDF and the Probability of a Specific Value**

The PDF and the PMF seem very similar concepts. You may ask yourself why the difference (one indicates directly the probability of a certain event, and the other must be integrated). Do they not tell the same exact thing? Do they not give the probability of a random variable to assume a specific value?

Although it is somewhat true, in the case of PDF things are more complicated than they seem. In fact, remember that the PDF is used in the case of a continuous variable. While in general $f(x)$ has a formula and can be calculated, it is meaningless to ask what the probability of a specific x is. Let me explain with an example.

Say you are looking for the probability of getting exactly π in a uniform $[0,10)$ random variable X. To do this you will select randomly the digit at each decimal position. You start from the first decimal place. You are looking at 1 in 10 possibilities (you need the 3). The probability of getting two digits (3.1) right is now 1/100 (you multiply the probabilities), and 3 correct decimals have a probability of 1/1000. For n decimal places the probability of getting exactly n digits of π is 10^{-n}. Now, let n go to infinity (you want to get exactly π remember), and as you would expect, the probability of getting exactly π tends to 0. This is the reason why, in the case of continuous random variables, one speaks of probability **density** function and not simple of probability function. For continuous distribution one uses the concept of cumulative distribution function (CDF, see Sect. 8.2.3) that gives the probability of a random variable to be smaller than a certain value (and not exactly **that** value).

Example 8.2.2: **PDF and Why Integration Is Needed**

To intuitively understand why the PDF for continuous distributions needs to be integrated, let us consider an example. Imagine you are measuring the exact height of a person. Height is a continuous variable because it can take any value within a range (e.g. between 150 and 200 cm). Now, if you wanted to find the probability that someone is **exactly** 175.000000000 cm tall, it would be almost impossible to do because, in a continuous distribution, the chance of any one specific value (like exactly 175.000000000 cm) is infinitesimally small, essentially zero. Instead of focusing on a specific value, you calculate the probability over a **range** of values (information that is more meaningful and more useful), like the probability that a person's height is between 170 and 180 cm. To do this, we use the PDF to describe how likely different heights are, and then we integrate the PDF over the range of interest (in this example between 170 and 180 cm). The area under the curve of the PDF between 170 and 180 cm gives us the total probability for that range.

8.2.3 Cumulative Distribution Function

The CDF of a random variable X, denoted by $F(x)$, is a function that gives the probability that X will take on a value less than x. Mathematically, the CDF is defined as
$$F(x) = P(X \leq x)$$
where $P(x)$ indicates the probability of the random variable to assume a specific value x or ranges of values $(X \leq x)$. The CDF provides a complete description of the probability distribution of a random variable and can be used to calculate the probabilities associated with different events that involve the random variable. In the case of continuous variables, $F(x)$ is given by

$$F(x) = \int_{-\infty}^{x} f(x)dx \tag{8.1}$$

Definition 8.2.4: **Cumulative Distribution Function**

CDF of a random variable X, denoted by $F(x)$, is a function that gives the probability that X will take on a value less than x. Mathematically, the CDF is defined as follows:
$$F(x) = P(X \leq x)$$
where $P(x)$ indicates the probability of the random variable to assume a specific value x or ranges of values $(X \leq x)$.

8.2.4 Expected Value and Variance

For completeness, I will also give you the definition of the expected value and variance in terms of distributions. Mathematically, the expectation of a random variable X is defined as follows for a continuous and for a discrete random variable.

$$\mathbb{E}(X) = \int_{\mathbb{R}} x f(x)dx \text{ for a continuous variable} \tag{8.2}$$

and

$$\mathbb{E}(X) = \sum_{x} x p(x)dx \text{ for a discrete variable} \tag{8.3}$$

where the sum is intended over all the possible values of X. For the variance we can write the formula

$$\text{Var}(X) = \mathbb{E}[(X - \mu)^2] \tag{8.4}$$

and calculate it with Eq. (8.2) or (8.3). You will see an example of using Eq. (8.2) or (8.3) in Sect. 8.3.

8.3 The Normal Distribution

We have looked at many general properties and definitions, and to make all this more concrete it is useful to observe how they apply to a real distribution. The best choice is, naturally, the normal distribution, possibly the best known, and most widely used of all.[2] It is used because of its unique properties and the natural phenomena it describes. Its importance comes from several key aspects.

1. **Ubiquity in natural phenomena**: Many natural and social phenomena follow a normal distribution making it a powerful tool for modelling and understanding a wide range of real-world data.
2. **Central limit theorem**: The central limit theorem (discussion of this theorem goes beyond the scope of this book) states that, under certain conditions, the sum of a large number of random variables, regardless of their distribution, will be approximately normally distributed. This makes most of the hypothesis testing methods usable (for more information on the central limit theorem see [2, 16]).
3. **Simplicity and mathematical convenience**: The normal distribution is mathematically tractable, making it easy to calculate probabilities and conduct statistical tests.
4. **Basis for other distributions**: Many other important distributions are related to the normal distribution, such as the chi-squared, t, and F distributions. These relationships extend the utility of the normal distribution in statistical modelling and hypothesis testing.
5. **Parameter estimation**: In statistics and machine learning, the normal distribution is often assumed for the underlying data. This assumption simplifies the estimation of model parameters and enables the use of techniques like maximum likelihood estimation.
6. **Error modelling**: In many statistical models, especially linear regression, errors are assumed to follow a normal distribution. This assumption facilitates the interpretation of model results and the construction of confidence intervals and hypothesis tests.
7. **Benchmark for comparison**: The normal distribution serves as a reference point for assessing the distribution of empirical data. Deviations from normality can indicate the presence of skewness, outliers, or other important features in the data.

---

[2] One distribution to rule them all.

8.3 The Normal Distribution

Example 8.3.1: **Normal Distribution**

Normal distribution (also known as Gaussian) can be found in a very large number of phenomena. Here are three examples.

Height of adult humans: The height distribution in a population of adult humans follows a Gaussian distribution. Although there may be slight variations due to factors like gender and ethnicity, the overall distribution tends to be approximately bell-shaped, with most individuals clustered around the mean height and fewer individuals at the extreme ends of the height spectrum.

Scores on standardised tests: Scores on standardised tests, such as IQ tests or college entrance exams such as SAT or ACT, exhibit a Gaussian distribution. This means that most test-takers score around the average (mean) score, with fewer individuals scoring at the lower and higher ends of the score range. The distribution of scores is typically symmetric around the mean.

Measurement errors: In many scientific experiments and measurements, the errors associated with the measurements often follow a Gaussian distribution. This is known as the Gaussian error or normal error. These errors can arise due to various factors such as instrument precision, environmental fluctuations, or human factors. The Gaussian distribution provides a mathematical framework for modelling and understanding the distribution of these errors.

The PDF of the normal distribution is given by

$$f(x) = \frac{1}{\sigma\sqrt{2\pi}} e^{-\frac{(x-\mu)^2}{2\sigma^2}} \tag{8.5}$$

where:

- x indicates the data.
- μ is the mean of the distribution (for a justification of this check Sect. 8.4).
- σ is the standard deviation of the distribution (for a justification of this check Sect. 8.4).

You can see an example of its shape in Fig. 8.1 for $\mu = 0$ and $\sigma = 1$. The normal distribution is symmetric around its mean, and its shape is characterised by its mean and standard deviation. The mean determines the location of the centre of the distribution, while the standard deviation determines the spread or width of the distribution. The CDF of the normal distribution, denoted by $\Phi(x)$, represents the probability that a random variable X is less than or equal to a given value x. It is given by

$$\Phi(x) = \int_{-\infty}^{x} f(t)\,dt$$

where $f(t)$ is the probability density function of the normal distribution (see Eq. (8.5)). Note that there is no exact formula for $\Phi(x)$ and numerical methods for integration must be used when numerical values are needed.

> **Definition 8.3.1: Normal Distribution**
>
> The normal distribution is described by a PDF given by
>
> $$f(x|\mu,\sigma) = \frac{1}{\sigma\sqrt{2\pi}} e^{-\frac{(x-\mu)^2}{2\sigma^2}} \qquad (8.6)$$
>
> where:
> - x indicates the data.
> - μ is the mean of the distribution.
> - σ is the standard deviation of the distribution.

8.4 ★ Mathematical Description of the Normal Distribution

In this section I will provide a more detailed mathematical analysis of the normal distribution. Let us start with some notation. The normal distribution is indicated with the symbol $\mathcal{N}$ and is characterised by two parameters: μ and σ. In this section, we will justify their meaning (spoiler alert, they are the mean and standard deviation, as we have mentioned without justification in the previous section). A random variable X that follows a normal distribution is indicated with

$$X \sim \mathcal{N}(\mu, \sigma^2) \qquad (8.7)$$

Note that the small wiggly symbol ($\sim$) indicates that the random variable X follows the distribution $\mathcal{N}(\mu, \sigma^2)$. We will see why the parameter σ is squared very soon. For the moment ignore this fact. The normal distribution density function, often indicated with $f(x|\mu, \sigma)$, is given by

$$f(x|\mu,\sigma) = \frac{1}{\sigma\sqrt{2\pi}} e^{-\frac{1}{2}\left(\frac{x-\mu}{\sigma}\right)^2} \qquad (8.8)$$

From Eq. (8.8) it should be immediately clear that it is symmetric in $x - \mu$ or in other words is centred at $x = \mu$. It goes to zero for $x \to \pm\infty$ and has a bell shape. You can see it for $\mu = 0$ and $\sigma = 1$ in Fig. 8.1. Note that $\mathcal{N}(0, 1)$ is called the **standard normal distribution**. The parameters μ and σ have the following meaning:

- μ: the mean of the data
- σ: the standard deviation of the data

The proof of this is given in Sect. 8.4.1, which is slightly more mathematically advanced and can be skipped if the reader is not so mathematically inclined.

8.4 ★ Mathematical Description of the Normal Distribution

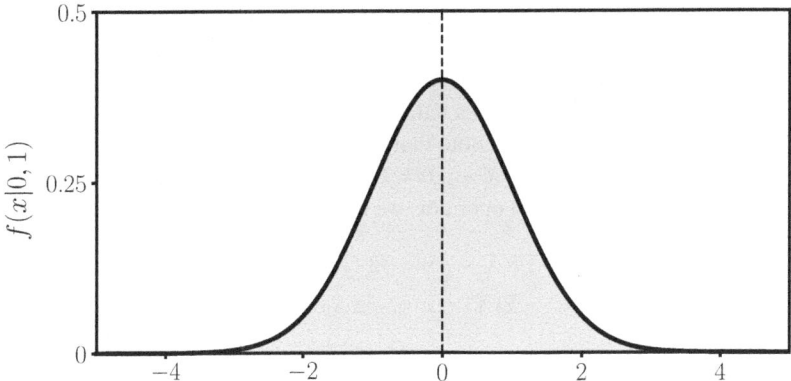

Fig. 8.1: The normal distribution density function for $\mu = 0$ and $\sigma = 1$. The normal distribution with those parameters is called a *standard* normal distribution

8.4.1 ★ Significance of μ and σ

Now, let us explore the significance of the parameters μ and σ. Let us start by calculating the expected value of a random variable that follows the normal distribution. We need to calculate

$$\mathbb{E}(X) = \int_{\mathbb{R}} x f(x|\mu, \sigma^2) = \int_{\mathbb{R}} x \frac{1}{\sigma\sqrt{2\pi}} e^{-\frac{1}{2}\left(\frac{x-\mu}{\sigma}\right)^2} dx =$$

$$= \frac{1}{\sigma\sqrt{2\pi}} \int_{\mathbb{R}} x e^{-\frac{1}{2}\left(\frac{x-\mu}{\sigma}\right)^2} dx = \left\{ s = \frac{x-\mu}{\sigma} \Rightarrow \sigma ds = dx \right\} =$$

$$= \frac{1}{\sqrt{2\pi}} \int_{\mathbb{R}} (\sigma s + \mu) e^{-\frac{1}{2}s^2} ds = \underbrace{\frac{\sigma}{\sqrt{2\pi}} \int_{\mathbb{R}} s e^{-\frac{1}{2}s^2} ds}_{A} + \qquad (8.9)$$

$$+ \underbrace{\frac{\mu}{\sqrt{2\pi}} \int_{\mathbb{R}} e^{-\frac{1}{2}s^2} ds}_{B}$$

Now note that the integral indicated with A is zero, given the symmetry of the function under the integral sign. Also note that we can use the known result.

$$\int_{\mathbb{R}} e^{-\frac{1}{2}s^2} = \sqrt{2\pi} \qquad (8.10)$$

to get the final result

$$\mathbb{E}(X) = \mu \qquad (8.11)$$

Tip 8.4.1: ★ Expected Value of a Symmetric Distribution

There is a very easy way to prove that for every random variable X that has a symmetric distribution around a value μ, it is true that $\mathbb{E}(X) = \mu$ regardless of the exact PDF. To prove it note that if X has a symmetric PDF around μ, then it must be true that $\mathbb{E}((X - \mu)) = \mathbb{E}(-(X - \mu))$; therefore, thanks to the linearity of the expectation operator, we have

$$\mathbb{E}((X - \mu)) = \mathbb{E}(-(X - \mu))$$
$$\mathbb{E}(X) - \mu = -\mathbb{E}(X) + \mu \quad (8.12)$$

and therefore by simply rearranging the terms, we have

$$\mathbb{E}(X) = \mu \quad (8.13)$$

That concludes the proof. That means that any random variable that has a symmetric PDF with respect to a value μ will have an expectation value (the average) equal to that value μ. This, of course, is true for the normal distribution, since it is symmetric around μ.

The parameter μ is the expected value of a random variable that follows a normal distribution.

Tip 8.4.2: ★ Proof of Eq. (8.10)

Let us define

$$A = \int_{\mathbb{R}} e^{-\frac{1}{2}s^2} \quad (8.14)$$

The trick to get the result in Eq. (8.10) is to calculate A^2 and not A. In fact we can write

$$A^2 = \int_{\mathbb{R}} dx \int_{\mathbb{R}} dy \, e^{-\frac{1}{2}x^2} e^{-\frac{1}{2}y^2} = \int_{\mathbb{R}} dx \int_{\mathbb{R}} dy \, e^{-\frac{1}{2}(x^2+y^2)} \quad (8.15)$$

Now we move to polar coordinates with the change of variables

$$\begin{cases} x &= r\cos(\theta) \\ y &= r\sin(\theta) \end{cases} \quad (8.16)$$

with r going from 0 to ∞ and θ going from 0 to 2π. Since we are making a change of variables (more than one), we need to calculate the Jacobian J. The reader should know that the following formula is valid:

$$dxdy = |J|drd\theta \quad (8.17)$$

8.4 ★ Mathematical Description of the Normal Distribution

where $|J|$ indicates the determinant of the Jacobian. In this case it is easy to see that

$$J = \begin{pmatrix} \frac{\partial x}{\partial r} & \frac{\partial x}{\partial \theta} \\ \frac{\partial y}{\partial r} & \frac{\partial y}{\partial \theta} \end{pmatrix} = \begin{pmatrix} \cos(\theta) & -r\sin(\theta) \\ \sin(\theta) & r\cos(\theta) \end{pmatrix} \quad (8.18)$$

therefore

$$|J| = r\cos^2(\theta) + r\sin^2(\theta) = r \quad (8.19)$$

With this result we can rewrite A^2 as

$$A^2 = \int_0^\infty dr \int_0^{2\pi} d\theta \, r e^{-\frac{1}{2}r^2} = 2\pi \int_0^\infty r e^{-\frac{1}{2}r^2} dr \quad (8.20)$$

which with the change of variable $s = r^2/2 \to ds = rdr$ can be easily calculated

$$A^2 = 2\pi \int_0^\infty e^{-s} ds = 2\pi \Rightarrow A = \sqrt{2\pi} \quad (8.21)$$

The reader should be able to do the integral in Eq. (8.21) easily. This concludes the proof.

To understand the parameter σ we need to calculate the variance of a random variable following a normal distribution. The integral to calculate is

$$\text{Var}(X) = \int_\mathbb{R} (x-\mu)^2 f(x|\mu, \sigma^2) = \int_\mathbb{R} (x-\mu)^2 \frac{1}{\sigma\sqrt{2\pi}} e^{-\frac{1}{2}\left(\frac{x-\mu}{\sigma}\right)^2} dx \quad (8.22)$$

since we now know that $\mathbb{E}(X) = \mu$. Let us evaluate the integral with the same change of variable we have done before, namely $s = (x-\mu)/\sigma$.

$$\text{Var}(X) = \int_\mathbb{R} s^2 \sigma^2 \frac{1}{\sigma\sqrt{2\pi}} e^{-\frac{1}{2}s^2} \sigma ds = \sigma^2 \underbrace{\frac{1}{\sqrt{2\pi}} \int_\mathbb{R} s^2 e^{-\frac{1}{2}s^2} ds}_{A} \quad (8.23)$$

We can show that $A = 1$. This tells us that the parameter σ^2 is nothing else than the variance of a random variable that follows a normal distribution:

$$\text{Var}(X) = \sigma^2 \quad (8.24)$$

Tip 8.4.3: ★ **Proof that $A = 1$ in Eq. (8.23)**

We want to prove that the following equation is valid:

$$A = \frac{1}{\sqrt{2\pi}} \int_{\mathbb{R}} x^2 e^{-\frac{1}{2}x^2} dx = 1 \qquad (8.25)$$

To do this we can use a neat trick. Let us start with the integral

$$I(\alpha) = \int_{\mathbb{R}} e^{-\alpha x^2} dx \qquad (8.26)$$

and note that, similar to what we have done in the Proof of Eq. (8.10), it is easy to show that

$$I(\alpha) = \int_{\mathbb{R}} e^{-\alpha x^2} dx = \sqrt{\frac{\pi}{\alpha}} \qquad (8.27)$$

Now we can take the derivative[a] of $I(\alpha)$

$$\frac{dI(\alpha)}{d\alpha} = \frac{d}{d\alpha} \int_{\mathbb{R}} e^{-\alpha x^2} dx = \int_{\mathbb{R}} \frac{d}{d\alpha} e^{-\alpha x^2} dx =$$
$$= -\int_{\mathbb{R}} x^2 e^{-\alpha x^2} dx \qquad (8.28)$$

But from Eq. (8.27) we know that

$$\frac{dI(\alpha)}{d\alpha} = -\frac{\sqrt{\pi}}{2} \alpha^{-3/2} \qquad (8.29)$$

by equating Eqs. (8.28) and (8.29) we get

$$\int_{\mathbb{R}} x^2 e^{-\alpha x^2} dx = \frac{\sqrt{\pi}}{2} \alpha^{-3/2} \qquad (8.30)$$

and by choosing $\alpha = 1/2$ we get the final result

$$\int_{\mathbb{R}} x^2 e^{-x^2/2} dx = \sqrt{2\pi} \qquad (8.31)$$

This concludes the proof.

---

[a] We will not discuss here the applicability of exchanging the derivation and the integral sign. This kind of discussion goes beyond the scope of the book.

8.4 ★ Mathematical Description of the Normal Distribution

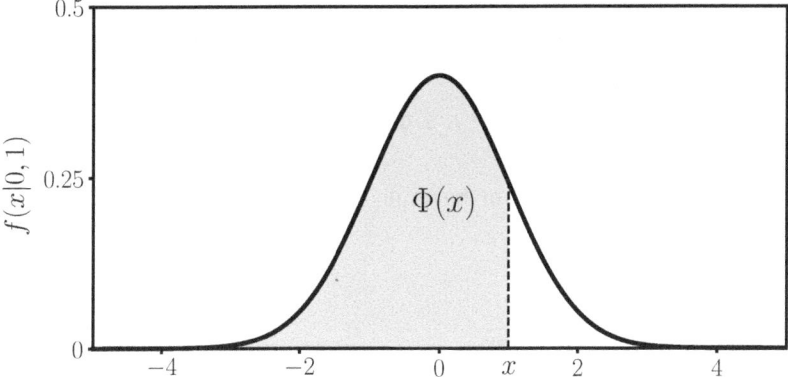

Fig. 8.2: $\Phi(x)$ is the value of the area greyed out. Note that $\Phi(x)$ is the symbol for the CDF for a standard normal distribution not for a generic normal distribution

Now let us calculate the CDF of a normal distribution. Let us start with the CDF of a standard normal distribution (remember that is a normal distribution with $\mu = 0$ and $\sigma = 1$). This is usually indicated with $\Phi(x)$ and is defined by

$$\Phi(x) = \int_{-\infty}^{x} f(x|0,1) = \frac{1}{\sqrt{2\pi}} \int_{-\infty}^{x} e^{-t^2/2} dt \qquad (8.32)$$

Note that we cannot express this integral in closed form. So in case you need to evaluate it, you need to do it numerically. $\Phi(x)$ is simply the area under $f(x|0, 1)$ from $-\infty$ to x, which you can see in Fig. 8.2 greyed out. The CDF of a general normal distribution can be expressed with $\Phi(x)$ as

Table 8.1: Relevant formulas related to the normal distribution

Function	Formula
$f(x\|0, 1)$	$\frac{1}{\sqrt{2\pi}} e^{-\frac{1}{2}x^2}$
$f(x\|\mu, \sigma)$	$\frac{1}{\sigma\sqrt{2\pi}} e^{-\frac{1}{2}\left(\frac{x-\mu}{\sigma}\right)^2}$
$\Phi(x)$ (CDF Standard Normal Distribution)	$\Phi(x) = \frac{1}{\sqrt{2\pi}} \int_{-\infty}^{x} e^{-t^2/2} dt$
$F(x)$ (CDF Normal Distribution)	$F(x) = \frac{1}{\sigma\sqrt{2\pi}} \int_{-\infty}^{x} e^{-\left(\frac{t-\mu}{\sigma}\right)^2/2} dt = \Phi\left(\frac{x-\mu}{\sigma}\right)$

$$F(x) = \frac{1}{\sigma\sqrt{2\pi}} \int_{-\infty}^{x} e^{-\left(\frac{t-\mu}{\sigma}\right)^2/2} dt = \left\{ s = \frac{t-\mu}{\sigma} \Rightarrow \sigma ds = dt \right\}$$
(8.33)
$$= \frac{1}{\sqrt{2\pi}} \int_{-\infty}^{\frac{x-\mu}{\sigma}} e^{-s^2/2} ds = \Phi\left(\frac{x-\mu}{\sigma}\right)$$

In Table 8.1 there is a summary of the results so far. The same results can be obtained for various other distributions.

8.5 Bernoulli Distribution

Let us start to define what a **Bernoulli** experiment is. A **Bernoulli experiment** is a type of random experiment where the only possible outcomes are two mutually exclusive results, typically classified as success or failure (e.g. flipping a coin resulting in heads or tails). When such an experiment is repeated multiple times **independently**, with the probability of success p remaining constant in each trial, it forms a **sequence of Bernoulli trials**. In this context, the probability of success is denoted by p, while $q = 1 - p$ represents the probability of failure.

> *Example* 8.5.1: **Bernoulli Trials**
>
> Consider this example adapted from [1]. Suppose that the probability of germination of a plant seed is 0.8, and the germination of a seed is called a success. If we plant ten seeds (we will assume that the germination of one seed is independent of another seed), this would correspond to ten Bernoulli trials with p = 0.8.

We can write the PMF for this type of random experiment. In fact, first of all, we know that the probability of $x = 1$ (success) is p. On the other hand, the probability of $x = 0$ (failure) is $q = 1 - p$. This can be conveniently written as one formula.

$$p(x) = p^x(1-p)^{1-x} = p^x q^{1-x} \text{ with } x = 0, 1 \quad (8.34)$$

In fact it is easy to verify that $p(0) = q$ and $p(1) = p$. Such a random variable X is said to have a **Bernoulli distribution**. The expected value (the mean) is given by

$$\mu = \mathbb{E}(X) = \sum_{x=0}^{1} x p^x (1-p)^{1-x} = (0)(1-p) + (1)p = p \quad (8.35)$$

8.5 Bernoulli Distribution

and its variance is given by

$$\sigma^2 = \text{Var}(X) = \sum_{x=0}^{1} (x-p)^2 p^x (1-p)^{1-x} \tag{8.36}$$

which turns out to be

$$\sigma^2 = \text{Var}(X) = (0-p)^2(1-p) + (1-p)^2 p = p(1-p) = pq \tag{8.37}$$

In a series of n Bernoulli trials, we use X_i to represent the Bernoulli random variable corresponding to the i-th trial. Thus, an observed sequence from n Bernoulli trials can be described as an n-tuple (a set of n values) consisting of zeros and ones, often called a random sample of size n from a Bernoulli distribution.

Example 8.5.2: **Bernoulli Distribution**

Here are three examples of random experiments that are described by a Bernoulli distribution.

Coin toss: The outcome of a single coin flip can be modelled as a Bernoulli random variable. You can, for example, define success as getting heads. Let the random variable X represent the outcome. We have

$$X = \begin{cases} 1 & \text{if heads (success)} \\ 0 & \text{if tails (failure)} \end{cases}$$

with probability $P(X = 1) = p$ and $P(X = 0) = 1 - p$ with $p = 0.5$.

Pass/Fail test: Consider a student taking a pass/fail exam. You can define success as passing the exam. Let the random variable Y represent the outcome. We have

$$Y = \begin{cases} 1 & \text{if pass (success)} \\ 0 & \text{if fail (failure)} \end{cases}$$

with probability $P(Y = 1) = p$ and $P(Y = 0) = 1 - p$, where p is the probability for a student to pass the test.

Defective item detection: In a quality control process, checking whether a randomly selected item from a production line is defective or not can be modelled as a Bernoulli random variable. Define success as finding a defective item. Let the random variable Z represent the outcome. We have

$$Z = \begin{cases} 1 & \text{if defective (success)} \\ 0 & \text{if non-defective (failure)} \end{cases}$$

with probability $P(Z = 1) = p$ and $P(Z = 0) = 1 - p$, where p is the probability that an item is defective.

8.6 Binomial Distribution

The binomial distribution is the probability distribution of the random variable that measures the number of successes in a fixed number of independent Bernoulli trials. If you perform n independent Bernoulli trials, each with the same success probability p, and count the number of successes X, then X follows a so-called **binomial distribution**.

To keep this section short, we will not give the derivations of the following functions. The PMF of a binomial distribution, where X is the total number of successes in n trials, is given by

$$P(X = k) = \binom{n}{k} p^k (1-p)^{n-k} \tag{8.38}$$

where:

- k represents the number of successes.
- $\binom{n}{k} = \dfrac{n!}{k!(n-k)!}$ is the binomial coefficient, representing the number of ways to choose k successes from n trials.
- p is the probability of success on a single trial.
- $n - k$ is the number of failures.

A binomial distribution describes the cumulative outcome based on repeated, independent trials, each trial being a simple 0/1, or success/failure situation represented by a Bernoulli random variable. It describes the outcomes of experiments involving multiple attempts with two possible outcomes per attempt, where each attempt is independent of the others. Its mean is np and its variance npq.

Example 8.6.1: **Bernoulli and Binomial Distribution**

This example illustrates the use of the Bernoulli distribution in a practical scenario: email spam detection. We consider each email as a trial with two possible outcomes: spam or not spam. Define a Bernoulli random variable X where:

- $x = 1$ indicates that an email is spam.
- $x = 0$ indicates that an email is not spam.

Suppose the probability of an email being spam is 0.2. Thus, we have $p = 0.2$ and $q = 1 - p = 0.8$. The email classification can be modelled by a Bernoulli distribution:

$$P(X = 1) = 0.2 \quad \text{(probability the email is spam)}$$

$$P(X = 0) = 0.8 \quad \text{(probability the email is not spam)}$$

If the company receives 100 emails per day, the number of spam emails is modelled by the **binomial distribution** with parameters $n = 100$ and $p = 0.2$:

- Expected number of spam emails per day: $E[X] = n \cdot p = 100 \cdot 0.2 = 20$.
- Variance of spam emails per day: $\text{Var}(X) = n \cdot p \cdot q = 100 \cdot 0.2 \cdot 0.8 = 16$.

8.7 The Poisson Distribution

The Poisson distribution models the number of events occurring within a fixed interval of time or space, assuming that these events occur with a constant mean rate and independently of the time since the last event. As examples consider traffic accidents (counting the number of traffic accidents that occur on a specific highway section each day), website hits (the number of hits on a website or views of a particular page within a given hour), or meteor shower observation (counting the number of meteors observed in a particular region of the sky during a meteor shower).

We can define a **Poisson process** as follows. Suppose we count the number of occurrences of some event in a given interval. This can be defined as an approximate **Poisson process** with parameter $\lambda > 0$ if the following conditions are satisfied:

1. The numbers of occurrences in non-overlapping intervals are independent.
2. The probability of one occurrence in a sufficiently short interval of length h is approximately λh.
3. The probability of two or more occurrences in a sufficiently short interval is zero.

Definition 8.7.1: **(Approximate) Poisson Process**

Suppose we count the number of occurrences of some event in a given interval. This can be defined as an approximate **Poisson process** with parameter $\lambda > 0$ if the following conditions are satisfied:

1. The numbers of occurrences in non-overlapping subintervals are independent.
2. The probability of one occurrence in a sufficiently short interval of length h is approximately λh.
3. The probability of two or more occurrences in a sufficiently short interval is basically zero.

Note that we use *approximate* to define the Poisson process since we use **approximately** in (2) and **basically** in (3) to avoid the "little o" notation (see Appendix B for a short introduction to the Big-O and little-o notation).

Note that we use *approximate* to define the Poisson process since we use **approximately** in (2) and **basically** in (3) to avoid the "little o" notation (see Appendix B for a short introduction to the Big-O and the little-o notation). This definition has been adapted from the wonderful discussion that can be found in Hogg et al. in [1].

We will skip the derivation of the formula here, as it would go beyond the scope of this short book, and we will redirect the reader to all derivations described in [1]. The PMF of a random variable X that counts the number of occurrences of events as we define above is given by

$$f(x) = \frac{\lambda^x e^{-\lambda}}{x!} \tag{8.39}$$

and we say that the random variable X has a **Poisson distribution**. Note that x indicates an integer here and not a continuous variable, since it indicates the number of occurrences. In addition, the mean and standard deviation of a random variable following a Poisson distribution are $\mu = \lambda$ and $\sigma^2 = \lambda$, respectively.

Example 8.7.1: **Poisson Distribution**

A tech support centre receives an average of 30 calls per hour. We want to find the probability of the centre receiving 25 calls in 1 hour. The probability of observing k events in an interval of time is given by the Poisson distribution formula:

$$P(X = k) = e^{-\lambda} \frac{\lambda^k}{k!} \tag{8.40}$$

where:

- λ is the average number of events per interval (30 calls/hour).
- k is the number of events of interest (25 calls).

Putting in the values, we calculate the probability of receiving exactly 25 calls in 1 hour

$$P(X = 25) = e^{-30} \frac{30^{25}}{25!} \approx 0.0511 \tag{8.41}$$

8.8 Probability Distributions: An Overview

The number of probability distributions existing and studied is really large, and this book is not the right place to describe them all. But for the sake of giving the reader an overview of the most famous, in Table 8.2 you can find an overview of the most famous distributions with a short description.

8.8 Probability Distributions: An Overview

Distribution	Description
Bernoulli distribution	The Bernoulli distribution is a discrete probability distribution that represents the outcome of a single trial, where there are only two possible outcomes: success or failure. It is characterised by a single parameter p, which is the probability of success. The distribution is the foundation for binary events, such as flipping a coin (heads or tails) or passing a test (pass or fail). See Sect. 8.5
Beta Distribution	The Beta distribution is a continuous probability distribution defined on the interval $[0, 1]$, with two shape parameters, α and β, that control the shape of the distribution. Depending on the values of these parameters, the Beta distribution can take on a variety of shapes, from uniform to U-shaped to bell-curved, making it highly versatile for modelling random variables that represent proportions or probabilities. It is especially useful in Bayesian statistics, where it often serves as a prior distribution for binomial data, such as the probability of success in Bernoulli trials. The flexibility of the Beta distribution allows it to reflect different levels of belief or uncertainty about a probability
Binomial Distribution	The Binomial distribution is a discrete probability distribution that models the number of successes in a fixed number of independent trials, where each trial has only two possible outcomes: success or failure. It is defined by two parameters: n, the number of trials, and p, the probability of success in each trial. Each trial follows a Bernoulli process, and the binomial distribution calculates the likelihood of obtaining a specific number of successes across all trials. The distribution is particularly useful for situations where the outcome of each trial is binary, and all trials are independent and identically distributed. See Sect. 8.6
Cauchy Distribution	The Cauchy distribution is a continuous probability distribution characterised by its heavy tails and undefined mean and variance. Unlike many distributions, it does not have finite moments of any order, making it unique in terms of its statistical properties. The Cauchy distribution is parameterized by a location parameter x_0, which defines the peak of the distribution, and a scale parameter γ, which determines the spread. It has a characteristic "bell shape" similar to the normal distribution, but with much heavier tails, meaning that extreme values are more likely to occur. The Cauchy distribution is often used to model data with significant outliers or when assumptions of finite variance do not hold
Chi-Square Distribution	The Chi-Square distribution is a continuous probability distribution that arises from the sum of the squares of k independent standard normal random variables. It is defined by a single parameter, k, which represents the degrees of freedom. The Chi-Square distribution is positively skewed (see Sect. 9.2), with the skewness decreasing as the degrees of freedom increase. It is commonly used in statistical inference, particularly in hypothesis testing (see Chap. 12) and constructing confidence intervals (see Chap. 11), especially in tests of goodness of fit. The distribution is also important in the analysis of variance (ANOVA) and for estimating population variances

Distribution	Description
Exponential Distribution	The Exponential distribution is a continuous probability distribution that models the time between events in a Poisson process, where events occur continuously and independently at a constant average rate. It is defined by a single parameter λ, which represents the average rate of occurrences. The Exponential distribution is *memory-less*, meaning that the probability of an event occurring in the future is independent of how much time has already passed. It is often used to model waiting times, such as the time between arrivals at a service point or the time until failure of mechanical systems
Gamma Distribution	The Gamma distribution is a continuous probability distribution that generalises the Exponential distribution and is defined by two parameters: a shape parameter α (often called k) and a rate parameter β (sometimes expressed as $\theta = 1/\beta$, the scale parameter). The distribution is used to model the time required for α independent events to occur in a Poisson process, making it a natural extension of the Exponential distribution when more than one event is involved. The Gamma distribution is versatile and used in various fields, particularly in queuing models, reliability analysis, and Bayesian statistics
Geometric Distribution	The Geometric distribution is a discrete probability distribution that models the number of trials needed to get the first success in a sequence of independent and identically distributed Bernoulli trials, where each trial has two possible outcomes: success or failure. It is characterised by a single parameter p, which represents the probability of success on each trial. The distribution is *memoryless*, meaning that the probability of success in future trials is independent of the number of failures that have occurred. The Geometric distribution is commonly used to model waiting times for the first occurrence of an event in repeated trials
Log-Normal Distribution	The Log-Normal distribution is a continuous probability distribution of a random variable whose logarithm is normally distributed. If a variable X is log-normally distributed, then $\log(X)$ follows a normal distribution. The Log-Normal distribution is defined by two parameters: the mean and standard deviation of the underlying normal distribution. It is commonly used to model non-negative data with a right-skewed distribution, such as income, stock prices, and certain biological measurements. The distribution is particularly useful when the data span several orders of magnitude
Negative Binomial Distribution	The Negative Binomial distribution is a discrete probability distribution that models the number of trials required to achieve a fixed number of successes in a sequence of independent and identically distributed Bernoulli trials. It is characterised by two parameters: r, the number of successes, and p, the probability of success in each trial. Unlike the Binomial distribution, which counts the number of successes in a fixed number of trials, the Negative Binomial distribution counts the number of trials needed to achieve the specified number of successes. It is often used to model count data, where the variance exceeds the mean

8.8 Probability Distributions: An Overview

Distribution	Description
Normal Distribution (Gaussian)	The Normal (Gaussian) distribution is a continuous probability distribution characterised by its symmetric, bell-shaped curve. It is defined by two parameters: the mean μ, which determines the location of the peak, and the standard deviation σ, which controls the spread of the distribution. The Normal distribution is widely used due to the central limit theorem, which states that the sum of a large number of independent random variables tends to follow a normal distribution, regardless of the original distribution. It is frequently applied in fields such as statistics, natural sciences, and finance to model data that clusters around a central value. See Sects. 8.3 and 8.4
Pareto Distribution	The Pareto distribution is a continuous probability distribution that models phenomena where large values are rare, but small values are common, following a power-law relationship. It is characterised by two parameters: the scale parameter x_m, which is the minimum possible value, and the shape parameter α, which governs the heaviness of the distribution's tail. The Pareto distribution is often used to describe the distribution of wealth, income, city sizes, and other systems where a small number of occurrences account for a large portion of the total effect
Poisson Distribution	The Poisson distribution is a discrete probability distribution that models the number of events occurring within a fixed interval of time or space, assuming the events occur independently and at a constant average rate. It is characterised by a single parameter λ, which represents the average number of events in the interval. The Poisson distribution is used to model rare events and is commonly applied in fields such as telecommunications, traffic flow, and biology to describe occurrences like the number of emails received in an hour or the number of mutations in a strand of DNA See Sect. 8.7.
Uniform Distribution	The Uniform distribution is a continuous or discrete probability distribution where all outcomes are equally likely within a specified range. For the continuous case, it is defined by two parameters, a and b, which represent the lower and upper bounds of the distribution, respectively. The probability density function is constant between a and b, meaning the probability of any outcome within this range is the same. The Uniform distribution is often used in simulations and random sampling where each outcome within a certain interval needs to be equally likely
Weibull Distribution	The Weibull distribution is a continuous probability distribution commonly used in reliability analysis and survival studies. It is characterised by two parameters: the shape parameter k and the scale parameter λ. The shape parameter controls the distribution's form, which can model various types of failure rates, including increasing, constant, or decreasing over time. The Weibull distribution is versatile and is widely applied in modelling lifetimes of products, systems, and materials, making it particularly useful in engineering and reliability testing

Table 8.2: An overview of the most famous probability distributions with a short description

Chapter 9
Skewness, Kurtosis, and Modality

9.1 Characteristics of a Distribution

It is always beneficial, when possible, to know the exact distribution that your data follow, as this allows you to infer many important properties of both the data and the phenomena being studied. However, in many cases, identifying the precise distribution is challenging or impossible. For instance, it is not immediately clear what distribution governs variables like maximum blood pressure, stock prices, cyberattacks, or solar flares.

When exact details are unavailable, a practical alternative for gaining insights into the phenomena is to examine the symmetry of the data distribution, the number of peaks it contains, and how quickly it decays at the extremes. We have previously covered concepts like quantiles, deciles, and percentiles, which help to capture asymmetries around the mean and median. While useful, there are more advanced tools for this purpose, which are the focus of this chapter.

Our goal is to have concise and easy-to-digest metrics that will give us information about the characteristics of a distribution. For the purposes for which we are interested in this chapter (asymmetry, number of peaks, and tail characteristics), we will discuss the concepts of skewness, kurtosis, and modality. Symmetry is measured by the **skewness** (see Sect. 9.2), number of peaks by the **modality** (see Sect. 9.4), and characteristics of the tails are assessed using **kurtosis** (see Sect. 9.3, where a nuanced discussion will be carried out).

Definition 9.1.1: **Tails of a Distribution**

The tails of a distribution are the portions of the distribution that can be found far left and far right of the mean or median.

> *Tip 9.1.1:* **Heavy and Light Tails (Intuitive Explanation)**
>
> The tails of a distribution are the portions of the distribution that can be found to the far left and far right of the central peak (imagine here a unimodal distribution). They are typically characterised by how quickly or slowly the probabilities decay as you move away from the mean or the mode of the data. The tails indicate the probability of finding extreme values in the data, either very high or very low, compared to the typical or average values represented near the peak(s) of the distribution.
>
> Tails are characterised into two types: **heavy** and **light**. Intuitively, a distribution is **light tailed** if large values are not so probable or **heavy tailed** if large values are more probable.

> *Definition 9.1.2:* ★ **Heavy Tail Distribution**
>
> A random variable X is said to have a heavy (right) tail distribution if
>
> $$\lim_{x \to \infty} \frac{P(X > x)}{e^{-\lambda x}} = \infty \qquad (9.1)$$
>
> for every real $\lambda > 0$. In other words, it means that the probability of finding the value of $X > x$ goes to zero slower than an exponential ($e^{-\lambda x}$).

Let us start with a discussion about how to measure the asymmetry of a distribution.

9.2 Skewness

Skewness is a measure of the asymmetry of the probability distribution of a real-valued random variable about its mean. The skewness (often indicated with γ_1) is defined as follows.

$$\text{Skewness} = \gamma_1 = \mathbb{E}\left[\left(\frac{X - \mu}{\sigma}\right)^3\right] \qquad (9.2)$$

where X is a random variable, μ is its mean, and σ is its standard deviation (assuming they exist of course).

9.2 Skewness

Definition 9.2.1: **Skewness**

Skewness is a measure of the asymmetry of the probability distribution of a real-valued random variable about its mean. The skewness (often indicated with γ_1) can be defined as follows:

$$\text{Skewness} = \gamma_1 = \mathbb{E}\left[\left(\frac{X-\mu}{\sigma}\right)^3\right] \tag{9.3}$$

where X is a random variable, μ is its mean, and σ is its standard deviation.

Let us discuss what it means when γ_1 has different signs. The sign and magnitude of the skewness indicate the direction and extent of the asymmetry.

- $\gamma_1 < 0$: The distribution has a longer left tail, which means it is concentrated towards the right. This condition is often referred to as left-skewed, left-tailed, or negatively skewed (see Fig. 9.1 for an example). Although the curve might appear to tilt to the right, "left-skewed" refers to the extended left tail and, typically, the mean being to the left of the median. Visually, such a distribution tends to lean towards the right.
- $\gamma_1 > 0$: The distribution has a longer right tail, where the bulk of the data are towards the left, which is known as right-skewed, right-tailed, or positively skewed (see Fig. 9.1 for an example). Even though the curve may seem to lean left, "right-skewed" indicates that the right tail is prolonged and the mean is generally to the right of median. Typically, this type of distribution appears to lean to the left.
- $\gamma_1 = 0$: This happens when the distribution is symmetric around the average (see Fig. 9.1 for an example).

Tip 9.2.1: ★ $\gamma_1 = 0$ **for a Symmetric Distribution**

It is easy to see that if the distribution is symmetric, $\gamma_1 = 0$. In fact let us consider a random variable with finite mean μ and finite standard deviation σ. Define a new random variable $Y = X - \mu$. We can write

$$\gamma_1 = \mathbb{E}\left[\left(\frac{X-\mu}{\sigma}\right)^3\right] = \frac{1}{\sigma^3}\mathbb{E}\left[(X-\mu)^3\right] = \frac{1}{\sigma^3}\mathbb{E}\left[Y^3\right] \tag{9.4}$$

Note that if the distribution of Y is symmetric around 0 (when X is symmetric around μ), then the distribution of Y and $-Y$ is the same. Thus we must have

$$\mathbb{E}\left[Y^3\right] = \mathbb{E}\left[-Y^3\right] = -\mathbb{E}\left[Y^3\right] \tag{9.5}$$

and this is only possible if $\mathbb{E}\left[Y^3\right] = 0$; thus $\gamma_1 = 0$.

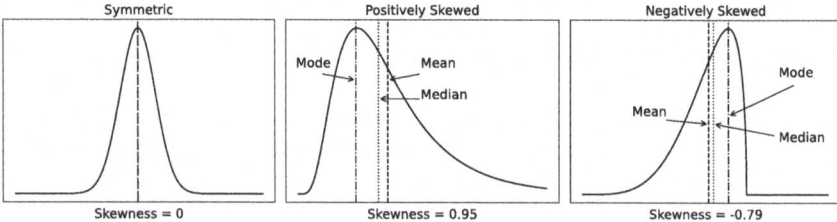

Fig. 9.1: Examples of distributions that are symmetric (left panel), positively (or right) (middle panel), and negatively (or left) (right panel) skewed, respectively. It is easy to see which is which: just look at the tail. If the longer tail is on the right, then is a right (positively) skewed distribution, and if the longer tail is on the left, then is a left (negatively) skewed distribution. The distribution in the left panel is perfectly symmetric, in fact it is a normal distribution, and thus the skewness is zero

You can see an example of distributions that are symmetric, positively (or right), and negatively (or left) skewed, respectively, in Fig. 9.1. It is easy to see which is which: just look at the tail. If the longer tail is on the right, then is a right (positively) skewed distribution, and if the longer tail is on the left, then is a left (negatively) skewed distribution.

> *Example* 9.2.1: **Right (Positively) Skewed Distributions**
>
> Right-skewed distributions are probably the most common. You can find them in data where there is a lower limit, and most of the data are close to this lower limit. Here are some examples.
>
> - Time to failure of a mechanical system or of a light bulb (for example) cannot be less than zero, but there is no upper bound.
> - The size of sales (say in USD) values is positive only, but the majority will be close to smaller values, with some exceptionally large sales.
> - Income is another good example. It is always positive, with the majority of values around the lower limit.

Now normally (**careful**: not always) in a right skewed distribution the mean is greater than the median and in a left-skewed is the opposite (again see an example in Fig. 9.1).

> *Example* 9.2.2: **Left (Negatively) Skewed Distributions**
>
> Left-skewed distributions are less common. Typically, you find them when the data have an upper limit, and most of the data values are close to the maximum.
>
> - Purity of substances cannot exceed 100%, and normally values are around this value.

9.2 Skewness

- Maximum test scores have a maximum, and typically students' test results are in the higher quartile.
- Ages at death is a perfect example. It has a clearly negatively skewed distribution.

Warning 9.2.1: **Relationship Between Mean and Median**

The relationship between skewness and the positions of the mean and median is not straightforward: A distribution with negative skew could have a mean that is greater than or less than the median, and the same applies to distributions with positive skew [17]. The assumption (and something you find in many textbooks) that mean, median, and mode are always in a given order in a skewed distribution is wrong, and it can fail, for example, in multimodal distributions or in distributions where one tail is long but the other is heavy.

Consider now another concrete example with the following PDF:

$$f(x, \mu, \sigma) = \frac{A}{x\sigma\sqrt{2\pi^2}} \cdot \exp\left(\frac{-(\log x - \mu)^2}{2\sigma^2}\right) \qquad (9.6)$$

In Fig. 9.2 you can see the distribution of 5000 points sampled from this PDF, where the factor $A \in \mathbb{R}$ is there to guarantee that the $f(x, \mu, \sigma)$ is normalised. In green you see the distribution for $\mu = 0, \sigma = 0.25$, in blue for $\mu = 0, \sigma = 0.5$, and in red for $\mu = 0, \sigma = 1$. You can see how the distributions (the PDFs, to be precise) are all asymmetric, with the red one being the most asymmetric and the green one the more symmetric. We can calculate the skewness γ_1 for the three cases (the values are coloured in Fig. 9.2), and we will get $\gamma_1 = 0.73$, $\gamma_1 = 1.72$, and $\gamma_1 = 2.16$. You can see how the more asymmetric the distribution is (the red one), the largest the value of γ_1 is. The closer γ_1 is to zero, the more symmetric the distribution is.

In Fig. 9.2 you can also see how the mean for the red PDFs is to the right of the median (for $\mu = 0, \sigma = 1$), which is highly asymmetric (**positively skewed distribution**).

9.2.1 Pearson's Skewness Coefficients

I would also like to discuss other ways of measuring skewness proposed by Pearson (known for the Pearson correlation coefficient). It is important to know them, since these may fail in some cases and thus must be used with the utmost care. Two coefficients are sometimes used: the **mode** coefficient and the **median** one. The **Pearson mode skewness coefficient** is defined by

$$\frac{\text{mean} - \text{mode}}{\text{standard deviation}} \qquad (9.7)$$

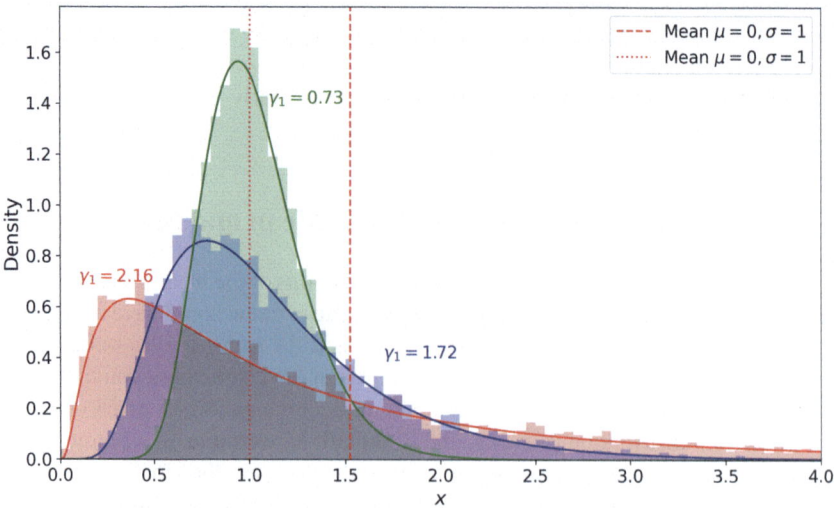

Fig. 9.2: Distribution of 5000 points sampled from the PDF in Eq. (9.6). In green you see the distribution for $\mu = 0, \sigma = 0.25$, in blue for $\mu = 0, \sigma = 0.5$, and in red for $\mu = 0, \sigma = 1$. You can see how the distributions (the PDFs, to be precise) are all asymmetric, with the red one being the most asymmetric and the green one the more symmetric. The skewness γ_1 for the three cases is printed in colour in the figure. The green one, the most symmetric, has (as expected) the lowest γ_1 value

Sometimes you find a slightly different definition of it, that is, the one above multiplied by 3. First of all, note that this is only usable when the distribution is unimodal. If you have multiple modes, you cannot use it, as it is not clear which mode you should use in the formula.

The **Pearson median skewness coefficient** is defined by

$$\frac{3(\text{mean} - \text{median})}{\text{standard deviation}} \tag{9.8}$$

Definition 9.2.2: **Pearson Mode and Median Skewness Coefficients**

The **Pearson mode skewness coefficient** is defined by

$$\frac{\text{mean} - \text{mode}}{\text{standard deviation}} \tag{9.9}$$

and the **Pearson median skewness coefficient** is defined by

$$\frac{3(\text{mean} - \text{median})}{\text{standard deviation}} \tag{9.10}$$

9.2 Skewness

From the example in Fig. 9.1 they seem a perfectly fine way of measuring how asymmetric a distribution is. The farther, for example, mean and median are (the more asymmetry there is), the larger the Pearson median skewness is. Unfortunately, this works in **most** of cases, but as we discussed it can happen that even in a negatively skewed distribution, you can have mean greater than median. Thus, the Pearson median skewness would suggest a wrong skewness. It is something to know and be aware of.

Warning 9.2.2: **Pearson Skewness Coefficients**

The Pearson skewness coefficients work in **most** of the cases, but it can happen that, even in a negatively skewed distribution, you can have the mean greater than the median for example. Thus, Pearson median skewness would suggest the wrong skewness sign. It is something to know and be aware of.

9.2.2 Quantile-Based Skewness Measures

There is an additional way to measure asymmetry that uses the quartiles that we discuss in Chap. 6. There are many possible variations, and we will only discuss one here. It is called Bowley's measure of skewness [18] or Yule's coefficient [19] and is given by the following definition:

$$\frac{Q_3 + Q_1 - 2Q_2}{Q_3 - Q1} \tag{9.11}$$

This can also be rewritten as

$$\frac{(Q_3 - Q_2) - (Q_2 - Q_1)}{Q_3 - Q1} \tag{9.12}$$

It is the difference of the range between the third quartile and the median and the range between the median and the first quartile divided by the IQR. If $(Q_3-Q2) > (Q_2-Q1)$ (positively skewed), you have a positive coefficient. If $(Q_3 - Q2) < (Q_2 - Q1)$ (negatively skewed), you get a negative coefficient.

9.2.3 Further Ways of Measuring Skewness

It is important to note that while there are many methods to measure skewness, the ones we have highlighted are the most well-known and widely used in statistics. These approaches should adequately meet any requirements and are suitable for almost all situations involving the analysis of distributions.

9.3 ★ Kurtosis

Kurtosis is another quantity that you may find in your statistical adventures. Its interpretation is not unambiguous and tends to be complex.

In general Kurtosis refers to the "tailedness" of a distribution. In general it is defined by the following formula:

$$\text{Kurtosis} = \mathbb{E}\left[\left(\frac{X-\mu}{\sigma}\right)^4\right] = \frac{\mathbb{E}[(X-\mu)^4]}{(\mathbb{E}[(x-\mu)^2])^2} \quad (9.13)$$

One of the most used symbols for kurtosis is κ, although there is no definite symbol that everyone uses. Often *excess kurtosis* is used instead, defined as

$$\text{Excess Kurtosis} = \mathbb{E}\left[\left(\frac{X-\mu}{\sigma}\right)^4\right] - 3 = \frac{\mathbb{E}[(X-\mu)^4]}{(\mathbb{E}[(x-\mu)^2])^2} - 3 \quad (9.14)$$

and the reason for the 3 comes from the normal distribution. In fact it can be shown that for a normal distribution

$$\text{Kurtosis(Normal Distribution)} = \mathbb{E}\left[\left(\frac{X-\mu}{\sigma}\right)^4\right] = \frac{\mathbb{E}[(X-\mu)^4]}{\sigma^4} = \frac{3\sigma^4}{\sigma^4} = 3 \quad (9.15)$$

Thus, *excess kurtosis* intuitively measures the kurtosis beyond that of a normal distribution (for which excess kurtosis is equal to zero).

> *Tip* 9.3.1: ★ **Proof Sketch that $\mathbb{E}[(X-\mu)^4] = 3\sigma^4$ for a Normal Distribution**
>
> This is a rough sketch on how one can show that for a normal distribution the kurtosis is equal to 3. Let $X \sim N(\mu, \sigma^2)$, meaning X is a normally distributed random variable with mean μ and variance σ^2. We start by standardising the random variable X. Define the random variable Z.
>
> $$Z = \frac{X-\mu}{\sigma}$$
>
> Since X is normally distributed, $Z \sim N(0, 1)$, meaning Z is a standard normal random variable with mean 0 and variance 1. Now, we want to compute $\mathbb{E}[(X-\mu)^4]$. Using the standardised variable Z, we have
>
> $$\mathbb{E}[(X-\mu)^4] = \mathbb{E}[\sigma^4 Z^4] = \sigma^4 \mathbb{E}[Z^4]$$

Thus, we need to compute $\mathbb{E}[Z^4]$, the fourth moment of the standard normal distribution. For a standard normal variable $Z \sim N(0, 1)$, the moments are known, and specifically
$$\mathbb{E}[Z^4] = 3$$
This result can be derived from the moment generating function of the standard normal distribution or using integration techniques. The exact calculation goes beyond the scope of this book (as it involves a complicated integration including the Gamma function) and thus will not be reported here. Substituting this into our expression, we get
$$\mathbb{E}[(X - \mu)^4] = \sigma^4 \cdot 3 = 3\sigma^4$$
Therefore, we have shown that
$$\mathbb{E}[(X - \mu)^4] = 3\sigma^4$$

□

As mentioned a normal distribution has an excess kurtosis of 0. When excess kurtosis is negative, the distribution is called **platykurtic**. This means that the distribution has fewer or less extreme outliers compared to a normal distribution. For example, the uniform distribution is platykurtic. In contrast, positive excess kurtosis indicates a **leptokurtic** distribution, where the tails decay more slowly than in a normal distribution, leading to more outliers. In general kurtosis (or excess kurtosis) assess the behaviour of the tails or in other words of values far from the mean. In fact the kurtosis measures the average (or expected value) of standardised data (value minus the average divided by the standard deviation) raised to the fourth power. That means that large values raised to the fourth power will have much more weight that values around the average. In fact values of $|X - \mu|/\sigma < 1$ (values distinct from the average less than one standard deviation) raised to the fourth power will become smaller. A number less than one raised to the fourth power will be much smaller than the original number, for example, $0.1^4 = 0.0001$. Analogously, values of $|X - \mu|/\sigma > 1$ (values farther than one standard deviation from the average) will become increasingly larger.

High values of excess kurtosis indicate that the distribution has heavy tails, meaning the probability of observing values far from the mean is greater compared to distributions with lower excess kurtosis.

9.4 Modality

Modality is used to describe the shape of a distribution based on the number of modes (peaks) it contains. There are four main types of modalities.

- **Unimodal**: a distribution with a single peak (e.g. the normal distribution). This is the most common distribution pattern, indicating a single most frequent value or range of values around a *central* point (note that the central point need not be the mean, as we have discussed in Sect. 9.2 on skewness).
- **Bimodal**: a distribution with two distinct peaks. This sometimes suggests the presence of two different processes or groups within the dataset.
- **Multimodal**: a distribution with more than two peaks. Multimodal distributions can indicate complex interactions within the data, with multiple groups or factors influencing the shape of the distribution.
- **Uniform**: a distribution where all values occur with roughly the same frequency, effectively showing no peaks.

> *Example* 9.4.1: **A Bimodal Distribution**
>
> A real-life example of a bimodal distribution can be found in the analysis of income distribution within an area where there are two dominant job sectors that differ significantly in average salaries. Consider a hypothetical city where there are high-paying jobs in a technology sector and low-paying jobs in the tourism sector. The technology sector offers high salaries, say ranging from 80,000 USD to 120,000 USD annually. In contrast, jobs in the tourism sector are lower paid, with typical salaries ranging from 20,000 USD to 40,000 USD. When analysing the overall income data for this area, you might see two peaks in the income distribution histogram: one peak around USD 30,000, representing the most probable salary in the tourism sector, and another peak around USD 100,000, representing typical earnings in the technology sector. This creates a bimodal distribution of income, with two distinct "modes" or peaks in the histogram. Each mode represents a cluster around a common value within the dataset, indicating that there are two typical income levels in the region, each associated with one of the two dominant sectors.

To analyse modality, one can look at histograms of the data to visualise the distributions and identify the number of peaks. Each mode often corresponds to a distinct subgroup within the dataset, making modality analysis valuable for segmenting data or identifying heterogeneous groups within the population. Understanding the modality of a dataset helps in understanding its underlying distribution and structure.

9.5 ★ Moments of a Distribution

You might be curious about how the definition of a concept such as skewness has that particular form. Where does it originate from? To answer this, we need to explore the concept of **moments** of a distribution. It is important to note that this topic is vast, and we will only scratch the surface here. This brief overview aims to illustrate how

9.5 ★ Moments of a Distribution

measures such as variance or skewness are mathematically connected and emerge naturally within the appropriate mathematical framework.

Tip 9.5.1: **Moments of a Distribution**

Moments of a distribution are fundamental statistical measures that provide insights into the shape and characteristics of a distribution. These moments, which include mean, variance, skewness, and kurtosis, each describe different aspects of the overall profile of the distribution.

Their name comes from the fact that they can all be derived by deriving one particular function, called **moment generating function** given by

$$M_X(t) = \mathbb{E}(e^{tX}) \tag{9.16}$$

The first moment is linked to the **mean**, the second moment to the **variance**, and the third and fourth moments to **skewness** and **kurtosis**, respectively. The beauty of moments lies in their ability to summarise complex datasets with simple, interpretable numbers that capture both central tendencies and variabilities. Thus, understanding moments is not just about dealing with abstract mathematical concepts; it is about gaining a more nuanced understanding of the data.

Consider X a random variable with CDF F_X and pdf f_X. The **moment generating function (mgf)** of X is denoted by $M_X(t)$ and is given by

$$M_X(t) = \mathbb{E}(e^{tX}) \tag{9.17}$$

if the expectation exists for t around 0. For a continuous X we have

$$M_X(t) = \int_{\mathbb{R}} e^{tx} f_X dx \tag{9.18}$$

and for a discrete X

$$M_X(t) = \sum_x e^{tx} p(x) \tag{9.19}$$

where $p(x)$ is the MDF of the discrete variable X, and the sum is intended over all values of X. We can also define the additional quantity

$$M_X^{(n)}(t) = \frac{d^n}{dt^n} M_X(t) \tag{9.20}$$

and this is called the n-th **moment** of the distribution. We can prove that

$$\mathbb{E}(X^n) = M_X^{(n)}(0) \tag{9.21}$$

Tip 9.5.2: ★ **Proof that** $\mathbb{E}(X^n) = M_X^{(n)}(0)$

To prove Eq. (9.21) we can simply follow the next steps.

$$M_X^{(n)}(t) = \frac{d^n}{dt^n}\mathbb{E}(e^{tX}) = \{\text{by using the Taylor expansion of the exponential}\}$$

$$= \frac{d^n}{dt^n}\mathbb{E}\left(\sum_{m=0}^{\infty} \frac{t^m X^m}{m!}\right)$$

$$= \frac{d^n}{dt^n}\sum_{m=0}^{\infty}\mathbb{E}\left(\frac{t^m X^m}{m!}\right) =$$

$$= \sum_{m=0}^{\infty}\frac{d^n}{dt^n}\left(\frac{t^m}{m!}\right)\mathbb{E}(X^m) =$$

$$= \sum_{m=n}^{\infty}\frac{m! t^{m-n}}{m!(m-n)!}\mathbb{E}(X^m)$$

$$= \frac{t^{n-n}}{(n-n)!}\mathbb{E}(X^m) + \sum_{m=n+1}^{\infty}\frac{t^{m-n}}{(m-n)!}\mathbb{E}(X^m)$$

$$= \mathbb{E}(X^m) + \sum_{m=n+1}^{\infty}\frac{t^{m-n}}{(m-n)!}\mathbb{E}(X^m) \quad (9.22)$$

By setting $t = 0$ we obtain the result.

From Eq. (9.21) is clear that the mean of X is the first moment, in fact

$$\mathbb{E}(X) = M_X^{(1)}(0) \quad (9.23)$$

Now it is easy to show that the variance can be written as

$$\sigma^2 = \mathbb{E}(X^2) - \mathbb{E}(X)^2 \quad (9.24)$$

so you can see how the variance is linked to two moments of the distributions, and in particular to the second moment. Of course, skewness, being linked to the expectation value of the third power of X, is naturally linked to the third moment of the distribution.

Definition 9.5.1: **Moment Generating Function (mgf)**

The **moment generating function (mgf)** of a random variable X with CDF F_X is denoted by $M_X(t)$ and is given as

$$M_X(t) = \mathbb{E}(e^{tX}) \qquad (9.25)$$

if the expectation exists for t around 0.

It is enough to rescale the random variable X, and suddenly mean, variance, skewness, and kurtosis will be exactly the first, second, third and fourth moments, respectively. Let us see how to do that.

9.6 ★ Central Moments

To make the relationship between moments and useful quantities (as the Variance) more direct, it is useful to define the central moments. Analogously as what we have done in the previous section, we can define a moment generating function

$$C_X(t) = \mathbb{E}(e^{t(X-\mu)}) = e^{-t\mu} M_X(t) \qquad (9.26)$$

where $\mu = \mathbb{E}(X)$. The nth central moment is defined by

$$C_X^{(n)}(0) = \left. \frac{d^n}{dt^n} C_X(t) \right|_{t=0} \qquad (9.27)$$

Now consider the first moment $C_X^{(1)}(0)$.

$$C_X^{(1)}(0) = \left. \frac{d}{dt} C_X(t) \right|_{t=0} = \left. \left(-\mu e^{-t\mu} M_X(t) + e^{-t\mu} M_X^{(1)}(t) \right) \right|_{t=0} = 0 \qquad (9.28)$$

In fact[1]

$$\left. -\mu e^{-t\mu} M_X(t) \right|_{t=0} = -\mu \qquad (9.29)$$

and

$$\left. e^{-t\mu} M_X^{(1)}(t) \right|_{t=0} = M_X^{(1)}(0) = \mu \qquad (9.30)$$

so we obtain Eq. (9.28). This makes sense since the random variable X is now centred on zero, and its mean (associated to the first moment) is now zero.

---

[1] Recall that $M_X(0) = 1$.

Things get interesting when calculating the second moment. In fact you have (you can start from the first moment that we calculated in Eq. (9.28))

$$C_X^{(2)}(t) = \frac{d}{dt}\left(-\mu e^{-t\mu} M_X(t) + e^{-t\mu} M_X^{(1)}(t)\right) \quad (9.31)$$

and with the previous calculation we can see that

$$C_X^{(2)}(t) = \underbrace{-\mu C_X^{(1)}(t)}_{A} + \frac{d}{dt}\left(e^{-t\mu} M_X^{(1)}(t)\right) \quad (9.32)$$

Now the term A in Eq. (9.32), when we set $t = 0$, will vanish, as we have proven above, so we can neglect it in our further calculations (recall that we are interested in calculating $C_X^{(2)}(0)$). Proceeding we get

$$\frac{d}{dt}\left(e^{-t\mu} M_X^{(1)}(t)\right) = -\mu e^{-t\mu} M_X^{(1)}(t) + e^{-t\mu} M_X^{(2)}(t) \quad (9.33)$$

Now if you check the previous equation, you may recall that

$$M_X^{(1)}(0) = \mu \quad (9.34)$$

and that

$$M_X^{(2)}(0) = \mathbb{E}(X^2) = \sigma^2 + \mu^2 \quad (9.35)$$

putting all together and setting $t = 0$, we finally get

$$C_X^{(2)}(0) = -\mu^2 + \sigma^2 + \mu^2 = \sigma^2 \quad (9.36)$$

and you can see how the second central moment is now exactly the variance of the random variable.

If you also consider the so-called *standardised* moments (meaning you do not only centralise X by subtracting the mean but also divide by the standard deviation), you will find out that the third standardised moment will turn out to be the skewness. Things are getting complicated and are out of scope for this book, so I will stop here.

My main goal with this advanced section is to show you how quantities used frequently in statistics that give you much information about a distribution (such as mean, variance, skewness, etc.) appear naturally as moments (raw, central, or standardised) of a distribution. Moments and the mgf are used in theoretical statistics to prove many results. For example, one of the simplest proofs of the central limit theorem[2] can be given with the mgf. This is a more advanced topic that we will not explore in more detail in this book. If you are interested, I suggest you read the book by Casella [2].

---

[2] The Central Limit Theorem intuitively states that if you take a large number of samples from any distribution and average them, the distribution of these averages tends to become a normal distribution, regardless of the shape of the original distribution.

Chapter 10
Data Visualisation

10.1 Histograms

A **histogram** is a type of bar graph that represents the distribution of numerical data by showing the frequency of data points within a certain range of values (called **bins**). To build a histogram you need to follow the steps below. Suppose you have a dataset $\{x_i\}_{i=1}^n$.

1. **Determine the Number of Bins:** Choose the number of bins k. Several approaches exist (we will discuss them later), but for the moment let us suppose that we choose k manually without any rule. For example, we could choose $k = 10$.
2. **Calculate the range of data:** Compute the range (see Sect. 5.3 for a discussion on the concept of *range*) of your dataset by subtracting the minimum value from the maximum value:

$$\text{range} = \max(x_i) - \min(x_i) \tag{10.1}$$

3. **Determine the bin width:** Divide the range by the number of bins k to find the bin width:

$$\text{Bin width} = \frac{\text{range}}{k} \tag{10.2}$$

4. **Create bin intervals:** Starting from the minimum data value, create bin intervals up to the maximum value, each with the calculated bin width. Your i-th interval Δ_i will be defined by

$$\Delta_i = \left[\min(x_i) + \frac{\text{range}}{k}(i-1), \min(x_i) + \frac{\text{range}}{k}i\right) \tag{10.3}$$

5. **Count the number of data points in each interval Δ_i (bin):** Count how many fall into each bin based on the established intervals.

© The Author(s), under exclusive license to Springer Nature Switzerland AG 2025
U. Michelucci, *Statistics for Scientists*,
https://doi.org/10.1007/978-3-031-78147-6_10

6. **Draw the histogram:** Draw the histogram with bins on the horizontal axis and the counts of data points on the vertical axis. The height of each bin corresponds to the number of data points it contains.

> *Example* 10.1.1: **Histogram of Test Scores**
>
> Suppose we have a dataset representing test scores from a class of 20 students:
>
> {55, 90, 68, 72, 83, 65, 88, 92, 78, 74, 69, 84, 77, 70, 90, 85, 89, 73, 67, 91}
>
> First, find the minimum and maximum values in your dataset. For the given data, the minimum score is 55, and the maximum score is 92. Decide how many bins (intervals) you want in your histogram. The choice may depend on the level of detail you wish to see. For simplicity, let us choose five bins. Subtract the smallest value from the largest value to find the range of your data. Then, divide this range by the number of bins to find the width of each bin.
>
> $$\text{Bin Width} = \frac{\text{Maximum} - \text{Minimum}}{\text{Number of Bins}} = \frac{92 - 55}{5} = 7.4$$
>
> Round this up to a convenient number, such as 8. Starting at the minimum value, add the bin width to create each bin interval (normally the interval includes the left limit but not the right one):
>
> - 55 to 63
> - 63 to 71
> - 71 to 79
> - 79 to 86
> - 86 to 94
>
> Count how many scores fall into each interval:
>
> - 55 to 63: 1 scores
> - 63 to 71: 5 scores
> - 71 to 79: 5 scores
> - 79 to 86: 3 scores
> - 86 to 94: 6 scores
>
> Using graph paper or any drawing tool:
>
> - Draw a horizontal line (*x*-axis) and label it with the bin intervals.
> - Draw a vertical line (*y*-axis) and label it with the frequency of scores.
> - For each bin, draw a bar that reaches the appropriate values.
>
> This histogram provides a visual representation of the distribution of scores, showing how they are spread across different intervals and helps to understand the overall performance trends in the class. See the resulting plot below.

10.1 Histograms

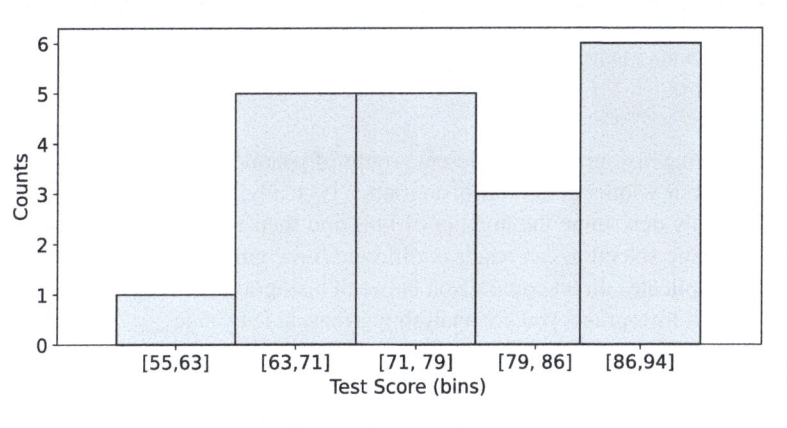

One decision you have to take is how many bins to use. This is an important decision since it can influence the histogram's effectiveness. Here are some rules that you can use to decide.

- **Sturges' Rule**: Number of bins = $\log_2 n + 1$, where n is the number of data points. Note that it may not provide enough bins for large datasets, leading to overly simplistic histograms that might miss important details in the data.
- **Square-Root Choice**: Use the square root of the number of data points.
- **Freedman-Diaconis Rule**: Bin width = $2 \times \text{IQR} \times n^{-1/3}$. It is more complex to calculate because of the need to calculate the interquartile range (IQR). It can also result in too many bins for large datasets, which can complicate the interpretation of the histogram.
- Adjust the bins **manually** to match the data distribution appropriately. In other words, you choose the width.

Tip 10.1.1: **Practical Tips for Building a Histogram**

When building a histogram, it is important to start by determining the purpose of the histogram, which will guide your choices throughout the process. Selecting the number of bins is the first step. You should begin with common rules like Sturges' or the square-root rule but always adjust based on the data to best reveal the distribution's shape (in other words, check the histogram visually). Consistency in bin width is fundamental, especially when comparing multiple histograms, as it ensures meaningful comparisons. Be careful of how outliers might affect the histogram's appearance, and consider adjusting the bin width. As a general rule, after generating the initial histogram, review it visually to ensure that it accurately represents the data. Finally, it is important to document (and publish together with your plots) your choices, including the method used to determine bin width and any adjustments made, to ensure reproducibility and clarity for others reviewing your work.

When creating histograms for different groups of your data, it is essential to maintain consistent bin widths across all histograms. Typically, Python packages or R will automatically determine the number of bins and their respective widths. However, this automatic selection can result in different bin counts for different data groups, which complicates direct comparison between histograms. Always set the same bin width for all histograms you are analysing groups side by side.

Tip 10.1.2: **Determination of the Number of Bins in Histograms**

After using a theoretical approach to determine the number of bins, visually inspect the histogram. Adjust the number of bins manually to see if a slight increase or decrease provides a more interpretable and insightful visualisation. The goal is to find a balance in which the shape of the data distribution is clear.

10.2 Boxplots

A **boxplot**, also known as a **box-and-whisker plot**, is a graphical representation of statistical data. In general, it serves the purpose of giving a condensed view of the distribution of the data by using graphically five numbers extracted from the data: (1) minimum (excluding outliers), (2) first quartile (Q1), (3) median (Q2), (4) third quartile (Q3), and (5) maximum (excluding outliers). The elements of a boxplot and a short discussion are reported below. You can see an example with all its elements in Fig. 10.1 where a boxplot is represented horizontally. Often those plots are vertical, but the elements are the same. We will consider Fig. 10.1 while discussing each part of a boxplot.

- **Box**: The main element of a boxplot is the central box that spans from Q1 to Q3 and represents how wide the central 50% of the data are, providing insights into the distribution's dispersion and skewness thanks to the median line (see later and additionally Sect. 9.2). For example, in Fig. 10.1 you can see how the 50% of the data goes from slightly above 9 to around 19.
- **Whiskers**: Whiskers are typically represented as horizontal lines (or vertical, depending on the orientation of the boxplot) that extend from the left and right (or bottom and top) of the box until the single points (the green points in Fig. 10.1). For example, the left whisker in Fig. 10.1 arrives at around 3, while the right one arrives slightly below 22.

10.2 Boxplots

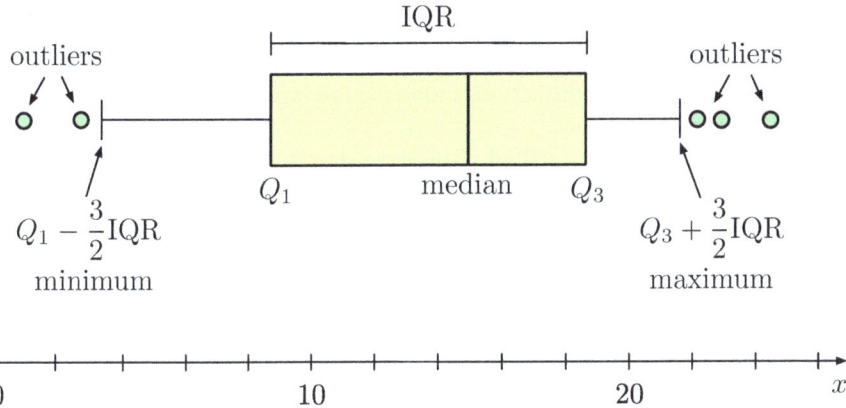

Fig. 10.1: The structure of a boxplot. The main elements that you will find in a boxplot are outliers, whiskers, the quartiles Q_1, Q_2 and Q_3, the IQR, and the median. The axis indicates the values of the data (indicated with x), and its values are only indicatives

- **Minimum**: The smallest data point **excluding** any outliers is represented by the end of the left whisker (when vertically orientated as the lower) or in Fig. 10.1 the small vertical lines that end the whisker lines and are at roughly around 3.
- **First Quartile (Q1)**: It represents the 25th percentile of the data. It is the median of the lower half of the dataset. It is identified in the boxplot by the left border of the box (or the bottom if the plot is orientated vertically) between 8 and 9 in Fig. 10.1.
- **Median (Q2)**: This is the middle value of the dataset, representing the 50th percentile. It is identified in the boxplot by a line across the box (not necessarily in the middle). In Fig. 10.1 we can see how the median is at around 15. From the box we can also see how the 25% of the data between the median and Q3 is more concentrated than the 235% of the data between the median and Q1.
- **Third Quartile (Q3)**: It represents the 75th percentile, marking the median of the upper half of the dataset. It is marked in the boxplot by the right border of the box (or the upper border if the plot is orientated vertically). In Fig. 10.1 Q3 is at around 19.
- **Maximum**: The largest data point **excluding** outliers is represented by the end of the right whisker (or the top one if orientated vertically). In Fig. 10.1 it can be found at around 22.
- **Interquartile Range (IQR)**: It is defined as Q3 minus Q1. This range covers the middle 50% of the data and is a measure of statistical dispersion. It is the width (or height) of the box.
- **Whiskers**: Lines that extend from Q1 to the minimum and from Q3 to the maximum (excluding outliers). They represent the spread of most of the data.

There are mainly two ways of deciding how long whiskers should be: Tukey's and the standard deviation method.

- **Tukey's method**: Whiskers extend to the last data point within 1.5 times the IQR from the quartiles.
- **Standard deviation method**: Whiskers extend to a specified number of standard deviations from the mean.

Not all boxplots are the same. Remember to always specify how you define the length of whiskers and therefore how you define outliers.

> *Tip* 10.2.1: **Outliers in Boxplots**
>
> Note that is good practice to always specify how you define the outliers in your boxplots. You can do that in the caption of the figure and in the text when you describe the figure. Not doing so will make unclear how you define outliers. Remember that.

- **Outliers**: Points that fall outside the range defined by the whiskers. They are plotted as individual points.

> *Warning* 10.2.1: **Limitations of Boxplots**
>
> The simplicity of boxplots also introduces certain limitations regarding the granularity of data they can display. Specifically, boxplots do not capture the finer details of a distribution's shape, such as its modality (see Sect. 9.4). Additionally, while boxplots show medians (see Sect. 4.3) and spread (see Chap. 5), they do not show the mode or mean explicitly (unless added additionally). They also do not depict the detailed distribution shape as histograms do.

You may remember our discussion about outliers in Chap. 7; now you see how such definitions are of practical use.

> *Tip* 10.2.2: **When to Use Boxplots**
>
> Boxplots are utilised to visualise the distribution of numerical data values and are particularly effective when comparing these distributions across various groups. They are designed to quickly convey high-level insights, providing an overview of a dataset's symmetry, skewness, variance, and outliers. They enable an easy assessment of the core concentration of data and facilitate comparisons between different groups.

10.3 Q-Q Plots

Suppose that you have a random variable X and have a hypothesis that it follows some distribution, for example, the normal one. How can you check if this is the case? And how can you visualise if the differences are in the tails or in the central parts? Q-Q plots address exactly this need. Q-Q stands for Quantiles-Quantiles. The basic idea is to compare experimental quantiles (coming from data) to theoretical ones (coming from the expected distribution) and plot one versus the other. If the data are distributed according to the expected distribution, all points will be on the diagonal line in the plot; otherwise they will be far from the diagonal.

> *Tip* 10.3.1: **Creating Q-Q Plots in Python**
>
> If you are interested in creating Q-Q plots in Python and in particular with matplotlib package, check the Python package probscale at https://matplotlib.org/mpl-probscale/.

Let us consider the *tips* dataset [20] that contains a column with tip amounts and the total amount of the bill for several people in restaurants. Let us consider the total_bill column, which, as the name suggests, contains the total amount paid. Let us first plot a histogram of the total_bill column to get an idea of the distribution of values. You can see it in Fig. 10.2. The distribution is asymmetrical and positively skewed (see Sect. 9.2). But we can ask ourselves how much it deviates from a normal distribution. To answer this question we can draw a Q-Q plot with the data quantiles versus the theoretical quantiles from the normal distribution. You can see the result in Fig. 10.3. Since the points (experimental quantiles) lie not on the line, we can say that the data are not distributed **perfectly** according to a normal distribution. But it is close, and you can see how the differences are not only in the tails (regions below 10 and above 40) but also, albeit in less amount, in the central part.

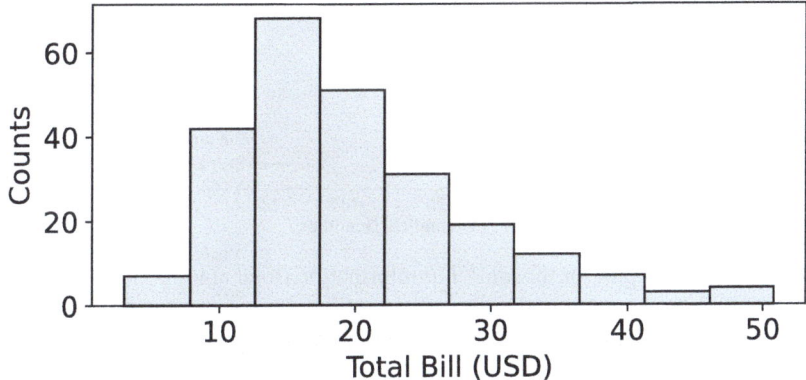

Fig. 10.2: Distribution of the column *total_bill* in the *tips* dataset

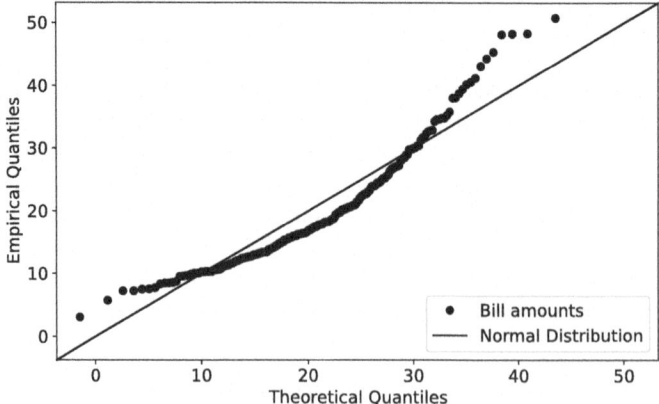

Fig. 10.3: The *Q-Q* plot for the *total_bill* information (total amount of bill) from the *tips* dataset [20] compared with a normal distribution. As you can see the data are not distributed normally, as the points deviate from the theoretical quantiles obtained from the normal distribution

How to find the best distribution that describes the data goes beyond the scope of this book, but by doing so you would discover that the data are best described by the Burr distribution (check [21] for more information). Check the *Q-Q* plot that compares the experimental quantiles with the theoretical ones coming from the Burr distribution in Fig. 10.4. You can see that the match is quite good, except for larger values, where apparently the data deviate from the Burr distribution.

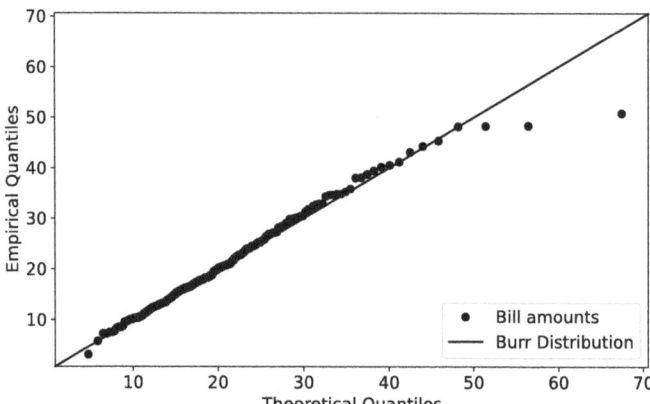

Fig. 10.4: The *Q-Q* plot for the *total_bill* information (total amount of bill) from the *tips* dataset [20] compared with a Burr distribution. As you can see the data follow the Burr distribution quite well, except for larger values

Tip 10.3.2: **Finding the Best Fitting Distribution in Python**

If you are interested in looking for the best distribution that fits your data, check the Python package `fitter` at https://fitter.readthedocs.io/en/latest/.

Q-Q plots are a nice visualisation tool to check visually how good a random variable follows a certain distribution.

10.4 Pair Plots

A pair plot displays all the pairwise relationships between variables in a dataset, along with the distribution of each variable. This is typically done in a grid where each row and each column represent one variable. Each cell in the grid contains a scatter plot of the corresponding pair of variables, while the diagonal cells often contain the univariate distribution of the variables, such as histograms or kernel density plots. For example, consider the Iris dataset.

Tip 10.4.1: **The Iris Dataset**

The dataset consists of 50 samples from three species of Iris: *Iris setosa*, *Iris virginica*, and *Iris versicolor*. For each species four characteristics were measured: length and width of the sepals and petals. In Fig. 10.5 you can see what petals and sepals are.

Fig. 10.5: Petals and sepals in a flower (Photo Eric Guinther CC BY-SA 3.0)

> The dataset comprises three categories, each with 50 samples, representing different types of iris plants. One category can be linearly distinguished from the other two, but the other two categories are not linearly distinguishable from one another. A nice article on the dataset that contains a lot of information is that by Unwin and Leinmann [22]. I suggest you read it, it is quite interesting.

In Fig. 10.6 you can see how such a plot looks like. Diagonal cells show the distribution of each variable. For instance, histograms or density plots can help you understand the spread and central tendency of each variable. Off-diagonal cells show

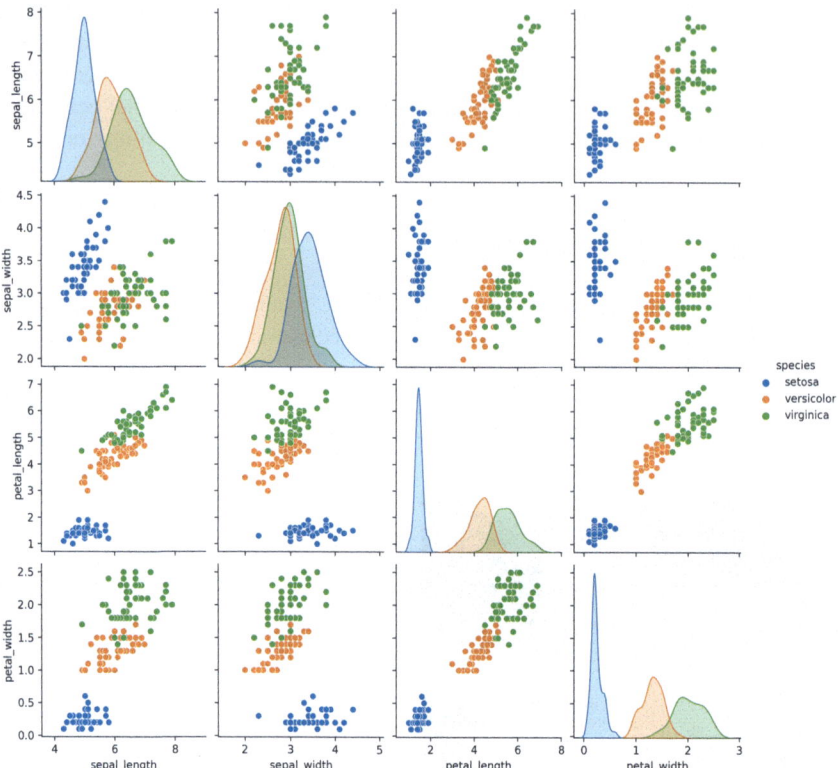

Fig. 10.6: A pair plot for the Iris dataset. For example, it is clear how `petal_length` and `petal_width` (search for the plot that is in the last column and second-to-last position from the top) are clearly correlated, as the points lie nicely along the diagonal of the plot. On the contrary, `sepal_length` and `petal_width` do not seem to be that correlated, as the points do not show any clear relationship (search for the plot that is in the first row, second position). You will get this pair plot by simply using the `seaborn` function `pairplot()`

10.4 Pair Plots

scatter plots for each pair of variables. For instance, a cell plotting `sepal_length` against `sepal_width` will show how these two measurements vary together. By looking at these cells, you can get an idea about possible correlations between variables. By using colour coding you can see how different categories (like species in the iris dataset) are distributed across the variable pairs in one shot. For example, by looking at Fig. 10.6, it is clear how `petal_length` and `petal_width` (search for the plot that is in the last column and second-to-last position from the top) are clearly correlated, as the points lie nicely along the diagonal of the plot. On the contrary, `sepal_length` and `petal_width` do not seem to be that correlated, as the points do not show any clear relationship (search for the plot that is in the first row, second position). For those of you interested, such a plot can be easily generated automatically in Python by using the `seaborn` library.

> *Tip* 10.4.2: **How to Read a Pair Plot and Why to Use It**
>
> Generally speaking, if you see a straight line (or almost a straight line) in a scatter plot, it indicates a linear relationship (correlation) between the variables. The slope of the line indicates the direction of the correlation (positive and negative). Different clusters in the scatter plots can indicate different groups or categories within the data. For instance, different species of flowers might form distinct clusters. Finally, points that are far away from others might be outliers and can be easily spotted in pair plots.
>
> Pair plots are typically used for Exploratory Data Analysis (EDA). They are extremely useful for getting a quick overview of the relationships between multiple variables in your dataset. They help in identifying linear or non-linear correlations between variables and patterns, for example, clusters, outliers, and trends in the data.

Chapter 11
Confidence Intervals

11.1 Introduction

Consider a random sample $X_1, X_2, \ldots, X_n$ drawn from a population that follows a normal distribution $\mathcal{N}(\mu, \sigma^2)$, where μ is the unknown population mean and σ^2 is the population variance. Our primary goal in statistics is to estimate the unknown population parameters (such as μ) based on the sample we have. Since it is usually impractical or impossible to measure the entire population, we rely on the sample to make educated guesses about the population's characteristics. Let us focus on estimating the population mean μ. A natural estimator for μ is the sample average

$$\bar{X} = \frac{1}{n} \sum_{i=1}^{n} X_i$$

which summarises the central tendency of the data in our sample (see Chap. 4). However, an important question remains: How close is the sample mean $\bar{X}$ to the true population mean μ. Since we are working with a finite sample, there will always be some degree of uncertainty or error in this estimate. This leads us to consider the variability of $\bar{X}$ as a random variable, which depends on both the sample size n and the inherent variability in the population (as captured by σ^2). Our goal is to quantify this uncertainty by constructing an interval that is likely to contain the true population mean with a certain level of confidence. This is the fundamental idea behind confidence intervals for the mean.

> *Example* 11.1.1: **Sample and Population Averages**
>
> Consider a standard normal distribution $\mathcal{N}(0, 1)$. If you sample ten values from this distribution, you will get something similar to (your numbers may be different, due to the random number seed and library you are

using) {1.39669393, 0.32622233, −0.90867756, 1.20833914, 1.71912777, 2.52668955, 0.53764895, −0.45544861, −0.61459241, −1.66651912}.
We know that the population average is $\mu = 0$, but if we calculate the average of our ten values, we will get $\bar{X} \approx 0.41$, far from zero. If we sample 10000 values and we calculate the average, we will get 0.001, a value much closer to zero. Thus, it is an important question of how close or far the sample average from $\mu = 0$ is.

11.2 Confidence Intervals for the Mean

Let us consider the average of a random variable X that follows a normal distribution. We can calculate the interval of X in which a certain percentage $1 - \alpha$ of values falls. α is called the confidence value. In other words, we want to find out a specific x_α such that

$$P\left(-x_{\alpha/2} \leq \bar{X} \leq x_{\alpha/2}\right) = 1 - \alpha \tag{11.1}$$

In this case, we do not have an analytical formula for $P(x)$, and thus it is typically not possible to calculate the value of $x_{\alpha/2}$ directly, since it would require the inversion of the function $P(x)$.[1] It is useful to write Eq. (11.1) in terms of $z = (\bar{X} - \mu)/(\sigma/\sqrt{n})$. We do this because a generic normal distribution with mean μ and variance σ^2 is transformed in a standard normal distribution with this change of variables. Let us continue with our calculation. We want to find the value $z_{\alpha/2}$ such that

$$P\left(-z_{\alpha/2} \leq \frac{\bar{X} - \mu}{\sigma/\sqrt{n}} \leq z_{\alpha/2}\right) = 1 - \alpha \tag{11.2}$$

with this notation we have

$$-z_{\alpha/2} \leq \frac{\bar{X} - \mu}{\sigma/\sqrt{n}} \leq z_{\alpha/2}$$

$$-z_{\alpha/2}\left(\frac{\sigma}{\sqrt{n}}\right) \leq \bar{X} - \mu \leq -z_{\alpha/2}\left(\frac{\sigma}{\sqrt{n}}\right)$$

$$-\bar{X} - z_{\alpha/2}\left(\frac{\sigma}{\sqrt{n}}\right) \leq -\mu \leq -\bar{X} + z_{\alpha/2}\left(\frac{\sigma}{\sqrt{n}}\right)$$

$$\bar{X} + z_{\alpha/2}\left(\frac{\sigma}{\sqrt{n}}\right) \geq \mu \geq \bar{X} - z_{\alpha/2}\left(\frac{\sigma}{\sqrt{n}}\right)$$

---

[1] Remember that the CDF for the normal distribution cannot be written in a closed analytical form, see Sect. 8.3.

11.2 Confidence Intervals for the Mean

which means we can rewrite Eq. (11.2) as

$$P\left(\bar{X} + z_{\alpha/2}\left(\frac{\sigma}{\sqrt{n}}\right) \geq \mu \geq \bar{X} - z_{\alpha/2}\left(\frac{\sigma}{\sqrt{n}}\right)\right) = 1 - \alpha \quad (11.3)$$

and we can interpret the equation by saying that *the probability that the interval*

$$\left[\bar{X} - z_{\alpha/2}\left(\frac{\sigma}{\sqrt{n}}\right), \bar{X} + z_{\alpha/2}\left(\frac{\sigma}{\sqrt{n}}\right)\right] \quad (11.4)$$

contains the probability mean μ is $1 - \alpha$. In other words once we have a sample and we calculate $\bar{X}$, we call the computed interval

$$\bar{X} \pm z_{\alpha/2}\left(\frac{\sigma}{\sqrt{n}}\right) \quad (11.5)$$

a $100(1 - \alpha)\%$ **confidence interval** for the unknown population mean μ.

Intuitively speaking, a confidence interval provides a range of values which is likely to contain the true value of an unknown population parameter. The interval has an associated confidence level that quantifies the level of confidence that the parameter lies within the interval (the $1 - \alpha$).

> *Tip* 11.2.1: **Meaning of Confidence Interval**
>
> It is important to give a precise interpretation of a confidence interval. For example, a 95% confidence level means that if you were to take N (with large N) different samples and calculate a confidence interval from each one, approximately 95% of those intervals would contain the true (population) average.

> *Example* 11.2.1: **Confidence Interval**
>
> Imagine you are trying to find the average height of all students in a large school. Instead of measuring every student, you randomly select a sample of students and calculate their average height. However, you know that the sample average may not be exactly the same as the true average height of all students in the school. To account for this, you can compute a confidence interval. In general if you take multiple samples from the same population, each sample will likely have a different average height due to random variability. A confidence level (e.g. 95%) indicates how certain you are that the true population parameter (e.g. the true average height) lies within the interval. For example, a 95% confidence level means that if you were to take 100 different samples and calculate a confidence interval from each one,

approximately 95% of those intervals would contain the true average height. The confidence interval is constructed around the sample mean and extends a certain number of standard errors (which measure the spread of the sample means) in both directions. If you calculate a 95% confidence interval for the average height to be between 160 and 170 cm, it means that you are 95% confident that the true average height of all students in the school is between 160 and 170 cm. Or, more correctly, a 95% confidence level means that if you were to take 100 different samples of students and calculate a confidence interval from each, approximately 95% of those intervals would contain the true average height.

Note that even with the well-known normal distribution, the evaluation of z_α requires numerical approaches. To evaluate z_α, which is the value corresponding to the desired confidence level for a standard normal distribution, you must rely on pre-calculated z-tables (you can find such tables in the book by Hogg, Tanis, and Zimmermann [1], for example, Table Va in Appendix B is for the standard normal distribution) or you have to invert (numerically) the cumulative distribution function (CDF) of the standard normal distribution. For example, For a 95% confidence interval, we have $\alpha = 0.05$ and $z_{0.025} \approx 1.96$. For a 99% confidence interval, we have $\alpha = 0.01$ and $z_{0.005} \approx 2.576$.

If the assumption that the random variable follows the normal distribution is not true, we can still get an approximate confidence interval for the mean μ. To assess how close (or far) a sample average from μ (the population mean) is, we can use the central limit theorem that tells us that

$$\bar{X} \sim N\left(\mu, \frac{\sigma^2}{n}\right) \tag{11.6}$$

Tip 11.2.2: **Central Limit Theorem**

The central limit theorem (CLT) is one of the most important results in probability theory and statistics. It provides a powerful way to make inferences about population parameters based on sample data. Intuitively, the CLT states that the distribution of the sample mean of a sufficiently large number of independent and identically distributed (i.i.d.) random variables approaches a normal distribution, regardless of the shape of the original distribution.

Consider a population with an unknown distribution and a random variable X measuring this population. Suppose that we take multiple samples of size n from this population. For each sample, we calculate the sample mean $\bar{X}$. We will assume that the samples are independent, which means that the selection of one sample does not influence the selection of another. They must also be identically distributed, meaning each sample is drawn from the same population distribution (another assumption we make). As the sample

11.2 Confidence Intervals for the Mean

size n increases, the distribution of the sample mean $\bar{X}$ approaches a normal distribution (this is the basic result given in the **central limit theorem**). This is true regardless of the shape of the original population distribution, as long as the original distribution has a finite mean μ and finite variance σ^2. The mean of the sample mean $\bar{X}$ will be equal to the population mean μ, and the standard deviation of the sample mean (often called the standard error) will be equal to the population standard deviation divided by the square root of the sample size n:

$$\mu_{\bar{X}} = \mu, \quad \sigma_{\bar{X}} = \frac{\sigma}{\sqrt{n}}$$

Imagine you have a large bag of jelly beans with different colours. You want to know the average number of red jelly beans in small handfuls taken from the bag. Each handful you take is a sample. If you keep taking more samples (handfuls) and calculate the average number of red jelly beans in each sample, the distribution of these sample averages will start to form a normal distribution (bell curve), even if the distribution of red jelly beans in the bag is not normal.

By the central limit theorem for large n the ratio $(\bar{X} - \mu)/(\sigma/\sqrt{n})$ follows an approximate (the larger n, the better the approximation) standard normal distribution, and thus we can apply the reasoning we described in this section.

Pre-calculated tables exist nonetheless for other distributions, most notably for the t-distribution, χ^2-distribution, and F-distribution, just to mention the most important ones. Nevertheless, if you are evaluating other statistical estimators (e.g. the median or other parameters in some statistical models), we cannot rely on the CLT or pre-calculated tables. In such a case, one method to evaluate confidence intervals is with the bootstrap approach (see Sect. 11.3).

Tip 11.2.3: **Confidence Intervals for Difference of Mean, Proportions and More**

There are ways of calculating confidence intervals for various estimators. For example, suppose you are studying, if two means are equal or different, in this case you would need confidence intervals for the different of the means (for more information check Hogg, Tanis, and Zimmermann's book in Section 7.2 [1]). You can even estimate confidence intervals for proportions, in cases where you are studying, for example, successes and failures proportion (for more information check Hogg, Tanis, and Zimmermann's book in Section 7.3 [1]). Many more cases have been studied, and depending on what you are doing is a good idea to take a good statistics book and check what is available. As you might have guessed by the references in this section, the book by Hogg et al. is a very good place to start.

11.3 Bootstrap Confidence Intervals

In general, as we have seen in the previous example, we need to know the distributions of the random variable X_i and the distribution of the statistical estimator (e.g. the mean) that we want to compute. If we do not have this information, we do not have at our disposal clear methods to estimate confidence intervals (e.g. in general we do not have pre-computed tables). To do this, we need to use a numerical approach that can be generally applied to many cases regardless of the underlying distributions. The most used and known is based on **bootstrap** resampling (first described by Efron in [9]).

> *Tip* 11.3.1: **Bootstrap in a Nutshell**
>
> Bootstrap is a resampling approach that allows you to create multiple samples from your data. It is, in its most basic form, quite simple. Starting from a sample $D = \{x_i\}_{i=1}^{N}$ of size N, a bootstrap sample consists in selecting N elements from D **with repetitions** (we discuss this resampling approach in Sect. 3.6). That means that some of the data points will appear multiple times in the bootstrap sample, but this method will allow you to generate multiple samples that you can use to study the statistical properties of an estimator.

Unlike traditional parametric methods that rely on specific distributional assumptions (such as normality), the bootstrap uses the sample data itself to approximate the sampling distribution. The basic steps to calculate confidence intervals using the bootstrap method are the following:

1. **Resampling:** From the original sample of size n, generate B *bootstrap samples* by randomly sampling, with replacement, n observations from the original data. Each bootstrap sample will have the same size n, but with some data points repeated and others omitted.
2. **Estimate the Statistic:** For each of the B bootstrap samples, compute the desired statistic (e.g. sample mean, median, etc.).
3. **Construct the Distribution:** The collection of B statistics forms the so-called *bootstrap distribution*, which approximates the sampling distribution of the statistic.
4. **Determine the Confidence Interval:** From the bootstrap distribution, calculate the desired confidence interval. This can be done numerically from the distribution obtained.

> *Example* 11.3.1: **Bootstrap for Confidence Interval Evaluation**
>
> Suppose we have a dataset of $n = 100$ observations, and we want to estimate a 95% confidence interval for the sample mean using the bootstrap. We follow these steps:

11.3 Bootstrap Confidence Intervals

- Generate $B = 1000$ bootstrap samples from the original data (with repetitions).
- Calculate the sample mean for each of the 1000 bootstrap samples.
- Sort the means and select the 2.5th and 97.5th percentiles from the bootstrap distribution to form the 95% confidence interval.

The bootstrap method is especially useful when the distribution is unknown. It provides a flexible and robust way to estimate confidence intervals without relying on strict parametric assumptions. However, it can be computationally intensive, especially with large datasets or complex statistics, as it requires generating many resamples and recalculating the statistic of interest.

The goal of this section was only to give you an idea about how bootstrap works, without the presumption of giving a proper treatment of the subject. For that I refer the reader to the book by Chernik [10] that offers a very good discussion of the method, its limitations, and applications.

Chapter 12
Hypothesis Testing

12.1 Disclaimer

Hypothesis testing is a slightly more advanced topic in statistics and requires the knowledge of more topics than what we have discussed in this short book. My goal is to explain, with a few examples, the main idea behind hypothesis testing. I will not go into all the different tests and how to use them, as there are many books that do that in a complete and clear way (see, e.g. [1]). My hope is that at the end of this chapter you will have a basic understanding of **how** hypothesis testing works and understand its basic principles. Let us start.

12.2 Hypothesis Testing: The Basic Idea

Consider the following example. Suppose that you are a trainer and you want to check if a new training regime will make your athletes faster. Let us indicate by X the time that an athlete can run 100 m. With the old training plan, you find, say, $\bar{X} = 12$ s, with $\bar{X}$ indicating the mean of X. After training with the new plan for a while, you find that $\bar{X} = 11.5$ s. How can we test whether the 0.5 s decrease in running time in the average is real or if it is only due to *luck* (maybe the wind was favourable) or, in other words, if it is not a real effect?

Such a test starts with a careful statement of the claims being compared. These claims are called in statistical terms **hypotheses**. The claim tested is called the **null hypothesis** and is often indicated by H_0. The test must be designed to assess the strength of evidence **against** the null hypothesis. The null hypothesis is tested against a competing claim, called the **alternative hypothesis** (and indicated with H_1 or H_a). Note that it is much easier to demonstrate that a statement is false. This is because proving something false requires only one counterexample, while proving something is true requires proving it in all possible situations. This is why we use a **null** hypothesis. The term null is used since this hypothesis claims that there is

no difference between two statistical measures (in our example, it would mean that the mean of the time needed to run 100 m is the same with the two training plans). Typically, the alternative hypothesis is the claim you are trying to prove true. Proving that H_a is true means proving that H_0 is false. Consider our example. As an example, our hypotheses could be:

- **Null Hypothesis** H_0: mean 100 m running time with athletes trained with the second plan $\mu = 12$ s
- **Alternative Hypothesis** H_1: mean 100 m running time with athletes trained with the second plan $\mu < 12$ s

The concept revolves around obtaining a numerical value (we will be more specific later) to assess the comparison between the two means and determine the probability (referred to as the *p*-value) of the truth or falsehood of H_0. Subsequently, by establishing a threshold for this probability that aligns with our acceptance criteria, we can determine the validity of each hypothesis.

12.3 An Example

Let us try to apply the method described intuitively to some real numbers. Consider a group of athletes again. Let us now imagine that the running times of athletes on a team that have trained in the past years are normally distributed, and $X \sim \mathcal{N}(12, 2.5)$ ($\mu = 12$ is measured in seconds and $\sigma^2 = 2.5$ in sec^2). We want to test a group of new athletes and check if they are slower on average or not. Suppose that the new group, say of eight athletes, has an average of $\bar{X} = 13$. We will assume that the standard deviation of athletes is known and is 2.5 s. That means that we will not use the values coming from the athlete samples, but the one we **know** to be true (maybe from scientific studies on athletes). This is an important point, as using the standard deviation estimated from the samples will make our testing more complicated. We will discuss this later in this chapter.

Now to check if the new athletes are slower, we need to ask the question of what is the probability of obtaining $\bar{X} = 13$ or greater when $\mu = 12$. This is called the *p*-**value** associated with $\bar{X} = 13$.

$$p\text{-Value} = P\left(\bar{X} > 13\right) = P\left(\frac{\bar{X} - 12}{\sqrt{2.5^2/8^2}} \leq \frac{13 - 12}{\sqrt{2.5^2/8^2}}; \mu = 12\right) \tag{12.1}$$

that is
$$p\text{-value} = 1 - \Phi(3.2) = 1 - 0.999 = 0.000687 \tag{12.2}$$

If this value is small (as it is in this example), we tend to reject the hypothesis that $\mu = 12$ for this new group of athletes. So, indeed, they are slower than the rest. In this example, we can write our hypotheses explicitly (with μ the average running time of 100 m of the new group).

12.3 An Example

- **Null Hypothesis** H_0: average 100 m running time of the new group of athletes $\mu = 12$ s
- **Alternative Hypothesis** H_1: average 100 m running time of the new group of athletes $\mu > 12$ s

For example, we could decide to reject H_0 if the *p*-value is below 0.05 (as it is said, with a confidence of 5%). Or, if we want to be really sure, if the *p*-value is below 0.01. This *p*-value is also indicated with α and can be interpreted as the probability of *rejecting H_0 when H_0 is true* (**type I** errors). Let us suppose that our alternative hypothesis is:

- **Alternative Hypothesis** H_1: average 100 m running time of the new group of athletes $\mu \neq 12$ s

Then we would need to calculate

$$p\text{-value} = P\left(\bar{X} > 13 \text{ and } \bar{X} \leq 13\right) = \tag{12.3}$$

which would result in two times the value we found above. Thus we would have

$$p\text{-value} = 0.001374 \tag{12.4}$$

Also in this case we would reject the hypothesis that the average of the new group is $\mu = 12$, or in other words we would accept H_1.

Tip 12.3.1: **Calculation of *p*-Value**

To understand Eq. (12.1) one needs to know that if $X_1, X_2, \ldots, X_n$ are observations of a random variable that distributed accordingly to $\mathcal{N}(\mu, \sigma^2)$, then the mean

$$\bar{X} = \frac{1}{n} \sum_{i=1}^{n} X_i \tag{12.5}$$

is distributed accordingly to $\mathcal{N}(\mu, \sigma^2/n)$. For a proof, you can check Corollary 5.5-1 in [1]. So when we want to calculate, for a random variable $X \sim \mathcal{N}(\mu, \sigma^2)$, say, the following probability

$$P(\bar{X} < c) \tag{12.6}$$

we can write

$$P(\bar{X} < c) = P\left(\frac{\bar{X} - \mu}{\sigma^2/n} < \frac{c - \mu}{\sigma^2/n}\right) \tag{12.7}$$

since we can scale both sides of the inequality. Now it is important to note that the random variable

$$\frac{\bar{X} - \mu}{\sigma^2/n} \sim \mathcal{N}(0, 1) \tag{12.8}$$

and thus is easy to calculate

$$P\left(\frac{\bar{X} - \mu}{\sigma^2/n} < \frac{c - \mu}{\sigma^2/n}\right) \qquad (12.9)$$

since we can even look it up in tables for the **standard** normal distributions. Note that this probability is the CDF of the **standard** normal distribution $\Phi(x)$ given by

$$\Phi(z) = \int_{-\infty}^{z} \frac{1}{\sqrt{2\pi}} e^{-\frac{x^2}{2}} dx \qquad (12.10)$$

12.4 Test of One Mean: Variance Known

In this case H_0 is typically of the form $H_0 : \mu = \mu_0$. There are three possibilities: (i) μ has increased, $H_1 : \mu > \mu_0$, (ii) μ has decreased, $H_1 : \mu < \mu_0$, and (iii) μ has changed, but we do not know in which direction, $H_1 : \mu \neq \mu_0$.

To test this, you would get a random sample of n observation and measure the mean $\bar{X}$. You will then assess the *closeness*, in terms of standard deviation of $\bar{X}$, σ^2/n. To measure this we will use the random variable

$$Z = \frac{\bar{X} - \mu_0}{\sqrt{\sigma^2/n}} = \frac{\bar{X} - \mu_0}{\sigma/\sqrt{n}} \qquad (12.11)$$

which is distributed according to the standard normal distribution. To test, for example, the hypothesis $\mu > \mu_0$, you would need to calculate

$$P(Z > Z_0) = p\text{-value} = P\left(\bar{X} > 13\right) = P\left(\frac{\bar{Z} - \mu_0}{\sqrt{\sigma^2/n^2}} > \frac{\bar{X} - \mu_0}{\sqrt{\sigma^2/n^2}}\right) \qquad (12.12)$$

and then reject or not H_0 according to the probability value. For a longer discussion, check Chapter 8 in [1].

12.5 Test of One Mean: Variance Unknown

We will need a random variable (like Z) that we can use to perform the same kind of analysis we have done in the previous section, but this time with the variances estimated from the samples. A natural choice is the variable (the one reason that we cannot explore here is that T is used to evaluate confidence intervals for means)

$$T = \frac{\bar{X} - \mu}{S/\sqrt{n}} \qquad (12.13)$$

12.5 Test of One Mean: Variance Unknown

where S^2 is the unbiased variance of the sample. The process works exactly in the same way as before, but to calculate the probabilities, we cannot use the normal distribution. In fact, it must be noted that the random variable T is not distributed normally, but accordingly to the so-called t-distribution. This is symmetrical and similar to the Gaussian (normal) distribution, but with heavier tails. Defining and discussing the t-distribution will go well beyond the scope of this short book. If you are interested in better understanding it, I suggest that you study Chapter 8 from [1]. Calculating the p-values in this case is as easy as before, thanks to many numerical software packages such as Python, SPSS, R, etc.

Warning 12.5.1: **Hypothesis Testing: More Complex Cases**

The types of test get more and more complicated the more assumptions you relax. For example, maybe you do not know the standard deviations, and in addition they are different in the two groups. The most important assumption in hypothesis testing is that the random variables are normally distributed. If this is not the case you have to use complex tests that do not use the normal distribution assumption. For example, here is a couple of tests that do not rely on the normality assumption to give you an idea.

- The **Wilcoxon Rank-Sum** Test, also known as the Mann–Whitney U Test, compares the medians of two independent samples to determine if they are drawn from the same population or if one tends to have larger values than the other. The test does not assume normality of the data but requires that the two samples have similar shapes of distributions.
- The **Kruskal–Wallis** Test is a non-parametric alternative to the one-way ANOVA test and is used to determine whether there are statistically significant differences between the medians of three or more independent groups. Like the Wilcoxon Rank-Sum Test, the Kruskal–Wallis Test does not require the assumption of normality but assumes that the shapes of the distributions of the groups are similar.

Tip 12.5.1: **Hypothesis Testing in Practice**

To summarise things a bit, let us sketch the process for doing hypothesis testing. Here is a short overview of the steps needed, described in an intuitive and somewhat superficial way, which I hope will help you understand the process.

1. **State the hypotheses:**
    - H_0: null hypothesis—the statement we want to test
    - H_a: alternative hypothesis—the opposite of the null hypothesis (what we want to test)

2. **Choose the significance level (α):**
   - Typically set to 0.05 (5%) or 0.01 (1%).
3. **Collect data:**
   - Gather relevant data through observation or experimentation.
4. **Calculate test statistic:**
   - Depending on the hypothesis being tested and the type of data collected, use the appropriate statistical formulae to compute the test statistic (e.g. Z or T). Some common test statistics include (but the list is far from exhaustive):
     - For comparing means: Independent Samples t-test, Paired Samples t-test, One-Way Analysis of Variance (ANOVA), or Wilcoxon Rank-Sum test
     - For comparing proportions: z-test for proportions or Fisher's Exact Test
     - For correlation: Pearson correlation coefficient or Spearman rank correlation coefficient for non-parametric data
     - For comparing variances: F-test or Levene's Test
5. **Determine critical value or p-value:**
   - Depending on the test statistic and the hypothesis being tested, find the critical value from the appropriate statistical table or calculate the p-value.
6. **Make a decision:**
   - If the test statistic falls in the critical region (reject region) or if the p-value is less than α, reject the null hypothesis.
   - If the test statistic falls outside the critical region or if the p-value is greater than α, fail to reject the null hypothesis.

12.6 p-Values: An Intuitive Definition

When hypothesis testing is performed, the p-value is, intuitively speaking, the probability of obtaining a result that is the one that was actually observed if H_0 is true. You will then **reject** the null hypothesis if the p-value is less than a predetermined value, often 0.05 (or 5%) or 0.01. Statistical software will give you p-values automatically, without needing to know how to calculate them, but it is important to be able to interpret the results correctly.

12.7 Type I and Type II Errors in Hypothesis Testing

It is useful to summarise some terminology on errors that are often used in statistical testing. A **Type I error** occurs when the null hypothesis (H_0) is true, but we incorrectly reject it.

> *Definition* 12.7.1: **Type I Error**
>
> A **Type I error** occurs when the null hypothesis (H_0) is true, but we incorrectly reject it.

This is also known as a "false positive" finding. The probability of committing a Type I error is often denoted by α, which is the level of significance used in the test. For example, if $\alpha = 0.05$, it means that there is a 5% risk of rejecting the null hypothesis even if it is true. A **Type II error** occurs when the null hypothesis (H_0) is false, but we incorrectly fail to reject it.

> *Definition* 12.7.2: **Type II Error**
>
> A **Type II error** occurs when the null hypothesis (H_0) is false, but we incorrectly fail to reject it.

This is known as a "false negative" finding. The probability of committing a Type II error is denoted by β. Type II errors are inversely related to the test power, which is defined as $1 - \beta$ (the probability of correctly rejecting a false H_0).

Chapter 13
Correlation and Linear Regression

13.1 Correlation

Correlation is a statistical measure that expresses the extent to which two variables are (linearly) related. It is a common tool used in statistics to analyse how one variable changes in relation to another. The most familiar measure of dependence between two quantities is the so-called **Pearson correlation coefficient**, denoted as r, which is a measure of the linear correlation between two variables X and Y. It has a value between +1 and −1, where 1 indicates total positive linear correlation (if one variable grows, the other does that too), 0 is no linear correlation (the two variables are not related in any form), and −1 is total negative linear correlation (if one variable grows, the other decreases). The formula for the Pearson correlation coefficient r between two variables x and y is given by

$$r = \frac{\sum_{i=1}^{n}(x_i - \bar{x})(y_i - \bar{y})}{\sqrt{\sum_{i=1}^{n}(x_i - \bar{x})^2}\sqrt{\sum_{i=1}^{n}(y_i - \bar{y})^2}} \tag{13.1}$$

where n is the number of pairs of values, and x_i and y_i are the individual sample points indexed with i. $\bar{x}$ is the mean of the x values, and $\bar{y}$ is the mean of the y values.

Definition 13.1.1: **Pearson Coefficient** r

Pearson correlation coefficient r between two variables x and y is defined by

$$r = \frac{\sum_{i=1}^{n}(x_i - \bar{x})(y_i - \bar{y})}{\sqrt{\sum_{i=1}^{n}(x_i - \bar{x})^2}\sqrt{\sum_{i=1}^{n}(y_i - \bar{y})^2}} \tag{13.2}$$

where n is the number of pairs of values, and x_i and y_i are the individual sample points indexed with i. $\bar{x}$ is the mean of the x values, and $\bar{y}$ is the mean of the y values.

A positive correlation indicates that as one variable increases, the other variable tends to also increase. A negative correlation indicates that as one variable increases, the other variable tends to decrease. Given Eq. (13.1) you may see how r can be expressed in terms of the standard deviations of the two quantities. Indeed r can be written as

$$r = \frac{\text{cov}(x, y)}{\sigma_x \sigma_y} \qquad (13.3)$$

where cov(x, y) is called covariance between x and y and is a measure of how two random variables vary together. It is defined as the expected value of the product of the deviations of two variables from their respective means. For two random variables x and y, with means $\bar{x}$ and $\bar{y}$, respectively, the covariance is given by the formula:

$$\text{cov}(X, Y) = \mathbb{E}[(x - \bar{x})(y - \bar{y})] \qquad (13.4)$$

where $\mathbb{E}$ denotes the expected value operator (if you do not remember what the expectation value is, check Sect. 4.2), or in another form

$$\text{cov}(X, Y) = \frac{1}{n} \sum_{i=1}^{n} (x_i - \bar{x})(y_i - \bar{y}) \qquad (13.5)$$

Definition 13.1.2: **Covariance**

The covariance between two random variables is defined by

$$\text{cov}(X, Y) = \mathbb{E}[(x - \bar{x})(y - \bar{y})] \qquad (13.6)$$

with means $\bar{x}$ and $\bar{y}$ of the random variables X and Y, respectively.

If the covariance is positive, it indicates that the two variables tend to increase or decrease together. In contrast, if it is negative, one variable tends to increase when the other decreases. A covariance of zero indicates that there is no linear relationship between the variables. Note, however, that zero covariance does not imply that the variables are independent unless they are jointly normally distributed (if you do not understand this point, do not worry about it).

13.1 Correlation

Warning 13.1.1: **The Pearson Coefficient and Non-linear Relationships**

The Pearson correlation coefficient measures the linear relationship between two variables. It is crucial to emphasise that it only captures the degree to which a **linear** relationship exists between the variables. If the relationship is **non-linear**, the Pearson coefficient may not accurately represent the strength of the association. In cases of non-linear relationships, other types of correlation coefficients, such as Spearman's rank correlation coefficient or Kendall's τ, might be more appropriate as they can capture monotonic relationships, whether linear or not.

That being said, it is still possible to compute the Pearson coefficient for any two variables, but one should be cautious in interpreting its value if the underlying relationship is known or observed to be non-linear. The coefficient might be low even if there is a strong non-linear relationship because Pearson's method is only looking for linearity.

To see why one must be careful, let us see an example. Suppose we generate points by adding some noise to the formula $y = x^2$. In Fig. 13.1 you can see the data with the red line that indicates the from which the data have been generated ($y = x^2$).

The black and the red functions are clearly strongly correlated, but if you calculate the Pearson coefficient, you will get -0.008. This would let you think that the data are not correlated, while the contrary is true. If you use, instead of the square function, $y = x$, the Pearson coefficient r will be 0.96 this time, indicating a strong correlation. So, to summarise, be careful when using statistical formulas without understanding them.

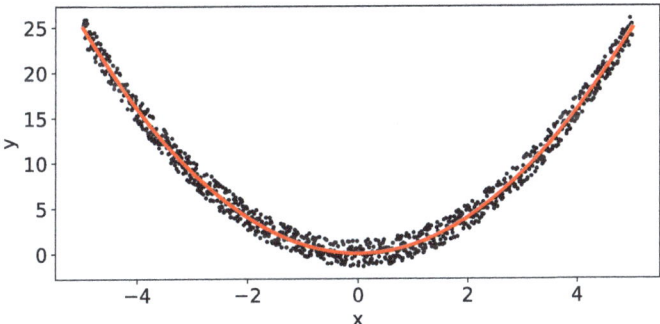

Fig. 13.1: The black points have been generated by adding random uniform noise with a magnitude of ± 3 to the red line (function $y = x^2$). If you evaluate r (the Pearson coefficient) between the black points and the red line, you will get -0.008 indicating no correlation, something that is clearly not true

There are many other ways of assessing correlation. Some of the most known are probably Spearman's Rank Correlation Coefficient (originally described in 1904 by Spearman in [23]) and Kendall's Tau (proposed in 1938 by Kendall in [24]) between others. If you need to assess correlation, be careful in which way it is done, and be aware of the limitations of the different methods.

13.2 Regression Analysis

Regression analysis is a statistical method used to model the relationship between a dependent variable (generally continuous) and one or more independent variables (called features). The main idea behind regression is to understand how changes in the independent variables influence the dependent variable, enabling us to make predictions, identify trends, and quantify the strength of relationships between variables. At its core, regression helps answer questions like: *If I change one variable, how does it affect another?* For instance, how does the number of hours studied affect exam scores? Or, in economics, how does the price of a product influence its demand? By fitting a mathematical model to the data, regression allows us to estimate the effects of one or more variables on the outcome.

The simplest form of regression is **linear regression**, where the relationship between the variables is modelled as a straight line. If we have only one independent variable, we call it *simple linear regression* (while when we have multiple independent variables, we call it *multiple linear regression*), and the model takes the following form:

$$y = \alpha_1 x + \alpha_0 + \varepsilon$$

where y is the dependent variable (what we are trying to predict, the outcome), x is the independent variable (the feature, or the input variable), α_0 is the intercept (the value of y when $x = 0$), α_1 is the slope of the line, and ε represents the random error or noise in the data.

> *Tip* 13.2.1: **Regression Beyond Linear Cases**
>
> Regression can be applied also in cases when the relationship between the independent and the dependent variables is not linear. This is called *non-linear* regression and is widely used in many cases.
> For example, non-linear regression is often used in biology to model population growth. Unlike linear relationships, biological systems frequently exhibit growth patterns that start slow, increase rapidly, and then level off as resources become limited. Another example is if we want to model the relationship between drug dosage and the concentration of the drug in the bloodstream over time. The absorption, distribution, metabolism, and excretion of drugs often follow non-linear patterns due to the complexities of biological systems.

13.2 Regression Analysis

At an intuitive level, regression provides us with a way not only to describe and explain relationships in the data but also to make predictions about future outcomes based on the patterns we discover.

13.2.1 Linear Regression

Very often, you will find yourself trying to understand **trends**. A trend (in the context of statistics) refers to the overall *direction* in which *something* tends to move over a period of time. It represents a pattern or general tendency of the data to increase, decrease, or remain stable.

> *Definition* 13.2.1: **Trend**
>
> A trend (in the context of statistics) refers to the overall *direction* in which *something* tends to move over a period of time. It represents a pattern or a general tendency of the data to increase, decrease, or remain stable.

For example, you may hear in the news that *the trend for buying new smartphones is flattening out*. This means that people buy fewer smartphones than before. This gives general and high-level information on how a certain phenomenon (in this example buying a new smartphone) changes over time and what to expect in the future. Generally, a trend is evaluated with *linear regression*.[1] Suppose you have some data (e.g. the number of smartphones that people buy every month in a specific country) that we will indicate with y_i (the i may indicate in our example the month). In our example, you would like to explore the relationship of y_i with time (the month). The y_i are called **dependent**, **outcome**, or **response** variable. The time in our example, which we can indicate with x_i, is often called the **independent**, **predictor**, or **explanatory** variable (often the x_i are called features in a machine learning language). In linear regression you assume that there is a linear relationship between the variables. In a more mathematical form

$$\hat{y}_i = \alpha_1 x_i + \alpha_0 \tag{13.7}$$

> *Tip* 13.2.2: **Regression**
>
> Regression is a method in statistical modelling that allows for the quantification of the relationship between a dependent variable (referred often as the outcome, target, or label) and one or more independent variables (known as predictors, features, or input features). This method is widely used to predict results, even when complex relationships exist.

---

[1] A trend can be assessed with other methods, but linear regression is the most widely used by far.

In general you may have more independent variables, which we can indicate with $x_i^{[j]}$ with $j = 1, \ldots, n$. In this case the linear relationship will assume the form

$$\hat{y}_i = \alpha_1 x_i^{[1]} + \alpha_2 x_i^{[2]} + \ldots + \alpha_n x_i^{[n]} + \alpha_0 = \sum_{j=1}^{n} \alpha_j x_i^{[j]} + \alpha_0 \qquad (13.8)$$

Our task is to find the α_i, known as **coefficients**, based on some criterion. Since we have some data (the y_i), we want to be able to find the coefficients α_i that will make the right side of Eq. (13.8) as close as possible to the expected values (the y_i). **Closeness** is measured in this case with the so-called Mean Squared Error (MSE) given by the formula

$$\text{MSE} = \frac{1}{N} \sum_{i=1}^{n} (y_i - \hat{y}_i)^2 \qquad (13.9)$$

where we have indicated with $\hat{y}_i$

$$\hat{y}_i = \alpha_1 x_i^{[1]} + \alpha_2 x_i^{[2]} + \ldots + \alpha_n x_i^{[n]} + \alpha_0 \qquad (13.10)$$

When you use statistical software (like Python or R), the coefficients that you will get are the ones that minimise the MSE.

Basically you want to find the coefficients α_j that minimise the MSE. This is what software packages and libraries do behind the scenes. Note that for linear regression we can solve the problem analytically. Meaning that we can write exact formulas for the coefficients that minimise the MSE, but that goes beyond the scope of this book. If you are interested in learning more about this, you can check Section 4.3.5 in my book *Fundamental Mathematical Concepts for Machine Learning in Science* [16], but to understand it you will need a solid foundation in linear algebra.

13.2.2 Coefficient of Determination

In the previous section, we have discussed how the determination of the parameters α_i works (albeit in a rather superficial and intuitive way). But it is important to have some way of assessing how "good" or "bad" a fit is. You can calculate such parameters for all kinds of data, even if they are not linear at all, so care is required in applying blindly linear regression.

To assess the "goodness" of the regression (sometimes also called *fit*) is by using the coefficient of determination that is typically denoted by R^2 or r^2. In what follows I will use the R^2 notation. Intuitively, R^2 is the proportion of the variation in the

13.2 Regression Analysis

dependent variable that is predictable from the independent variable(s). To calculate it we need two quantities: SS_{res} and SS_{tot}.

$$SS_{res} = \sum_{i=1}^{N}(y_i - \hat{y}_i)^2$$
$$SS_{tot} = \sum_{i=1}^{N}(y_i - \bar{y})^2 \tag{13.11}$$

where

$$\bar{y} = \frac{1}{N}\sum_{i=1}^{N} y_i \tag{13.12}$$

is the average of the y_i. Then R^2 is given by

$$R^2 = 1 - \frac{SS_{res}}{SS_{tot}} \tag{13.13}$$

and this can rewritten as

$$R^2 = 1 - \frac{SS_{res}}{SS_{tot}} = \frac{SS_{reg}}{SS_{tot}} \tag{13.14}$$

where we have written

$$SS_{reg} = SS_{tot} - SS_{res} \tag{13.15}$$

Equation (13.14) can be interpreted as the ratio of the variance that the regression explains (SS_{reg}) to the total variance of the data (SS_{tot}). The formal mathematical proof goes beyond the level of this book, and thus I will not report it here in this section. An intuitive understanding of the meaning of R^2 will serve you well. If you are interested in knowing how to prove it, you can check Appendix A.

Tip 13.2.3: R^2 Use Tips

R^2 represents the proportion of the variance in the dependent variable (response) that is explained by the independent variable(s) in the model. Values range from 0 to 1. When $R^2 = 0$, the model explains none of the variability in the data. When $R^2 = 1$, the model explains all of the variability in the data. For example, if $R^2 = 0.8$, it means 80% of the variation in the dependent variable is explained by the model, while 20% remains unexplained. In general a high value for R^2 indicates that a linear fit works well, but that does not necessarily mean that a linear regression is the right model! Keep that in mind.
R^2 assumes a linear relationship between the variables. If your R^2 is low, it might indicate that the relationship is non-linear. In such cases, consider

> applying transformations to your variables or using non-linear models to model the relationship more accurately.
>
> Finally, R^2 gives a sense of how well the model explains the data, but it does not provide information on prediction accuracy. For predictive performance, use metrics like Mean Squared Error (MSE), Root Mean Squared Error (RMSE), or Mean Absolute Error (MAE). Additionally it is very important to combine R^2 with visual checks (e.g. residual plots) to diagnose model performance more precisely.

13.3 Further Readings

Generally speaking, regression is a way of modelling the dependence of a continuous variable from some independent features (one or often many) (not only in the linear case). The most known method is linear regression, but it is possible to model any kind of relationships with various methods. If you are interested in learning more about regression, the famous book by Hastie et al. [25] (the version cited is the one with examples in Python, but in case you are interested you can find the second edition of it with code in R) is a great place to start. A lesser-known but highly comprehensive book is *Elements of Statistical Learning* by Hastie et al. [26], which provides an in-depth exploration of regression problems. Chapter 12 in Casella and Berger's book *Statistical Inference* [2] is also worth checking.

Chapter 14
Ethics and Best Practices

14.1 Steps of a Statistical Project

The following list contains seven steps that are needed in almost all statistical projects. In it, I tried to give some tips and hints to help beginners identify the most important aspects and challenges to pay attention to.

1. **Write one or multiple research questions**

   What to do: Start by brainstorming about what you want to know or understand better within your field of interest. Consider what questions have not been fully answered or what areas need further exploration. To do this do a literature research (a very good tool to do that is Google Scholar at https://scholar.google.com/) on the topic you are interested in, study the papers, and try to identify relevant questions. This is not easy, especially if you are at the beginning of your scientific career, and help from a more senior profile (postdoc or professor) will probably be necessary. See Chap. 3 for more details.

   Hints: Keep your research questions focused and specific. Avoid overly broad questions that are difficult to address within a single study. At the beginning it is very useful to reproduce some of the main results existing in literature if possible or at least try to get your hands on existing datasets and verify some of the results. This will give you a better understanding of the data you might need and the analysis you will need to do.

2. **Design multiple hypotheses**

   What to do: Begin by reviewing the existing literature related to your research questions. This can help you identify potential hypotheses based on previous findings or theories. Formulate hypotheses (see Chap. 3) that can be disproved with experiments. Make them as concrete as possible. Remember a hypothesis is an educated guess of a possible outcome of an experiment.

   Hints: Make sure that your hypotheses are testable and have clear predictions about the relationship between variables. Consider both null and alternative hypotheses to cover different possibilities. Avoid vague terms, and be precise about the variables and the expected relationship between them. Remember that your hypotheses must be disprovable with experiments, so clearly define the independent variable (the factor you manipulate or observe) and the dependent variable (the outcome you measure). Finally a good hypothesis should be able to be **proven false**. It should be possible to collect data that could potentially **refute** the hypothesis.

3. **Data Collection**

   What to do: Define your target population carefully, considering who or what you want your research findings to generalise to. Plan your data collection strategy in advance, including how you will collect data (e.g. surveys, experiments, and observations) and any tools or instruments you will use.

   Hints: Choose a sampling approach (see Chap. 3) that best suits your research objectives and resources. Random sampling is often preferred for its ability to produce representative samples, but other methods may be appropriate depending on the study design.

4. **Design the experiments**

   What to do: Clearly identify the variables you will manipulate (independent variables) and measure (dependent variables) in your experiment according to your hypotheses. Pilot test your experimental design before conducting the main study to identify and address any potential issues or limitations. If you are collecting data from people with surveys, try them with friends or some test subject. You will discover what does not work and how to fix it before wasting lots of time and money.

   Hints: Consider potential confounding variables that could affect your results and plan ways to control for them during the experiment.

> *Warning* 14.1.1: **Cofounding Variables**
>
> **Confounding variables** are variables (sometime not even measured) that can influence both the dependent variable and independent variable at the same time, potentially leading to a spurious association or masking a real association between the variables being studied.
>
> Consider a study investigating the relationship between physical activity (independent variable) and heart disease (dependent variable). Socioeconomic status (SES) could be a confounding variable if SES is associated with the level of physical activity (e.g. people with higher SES may have more access to gyms and time for exercise). SES also influences the risk of heart disease (e.g. people with higher SES may have better access to healthcare and healthier lifestyles). In this case, failing to account for SES could lead to incorrect conclusions about the relationship between physical activity and heart disease.

5. **Collect Data**

    What to do: Follow your data collection plan closely to ensure consistency and reliability in your data. Consider the ethical implications of your data collection methods, especially when working with human subjects, and ensure that you obtain informed consent when necessary (in written form if necessary). Note that this may take months and will make your study more challenging but is a necessary step when dealing with human subjects.

    Hints: Keep detailed records of your data collection process, including any unexpected observations or deviations from the plan. Remember we mentioned the notebook? Always keep **every** information in your notebook, positive, negative, and wrong. And do not forget to put a date on your notes.

6. **Analysis**

    What to do: Do the analysis you planned. Describe first the data with measures of central tendency (see Chap. 4), dispersion (see Chap. 5), and position (see Chap. 6). Use visualisation to show your data and to better understand its distribution.

7. **If applicable, do hypothesis testing (inferential statistics)**

    What to do: Choose appropriate statistical tests based on the nature of your hypotheses and the type of data you have collected. Interpret the results of your hypothesis tests carefully, considering both statistical significance and practical significance in the context of your research questions.

    Hints: Familiarise yourself with the assumptions underlying the statistical tests you plan to use and check whether they are met by your data (see Chap. 12 for a discussion).

In Table 14.1 you will find a short summary and overview of the steps that you can follow in any project with a statistical focus. Of course this list is not applicable to all projects, but it is a good start that will help you in structuring your work.

14.2 Reproducibility, Replicability, Transparent Reporting, and Documentation

Transparent reporting and documentation are critical components of scientific research that ensure that the findings are credible, reproducible, and verifiable. One of the main goals of any scientific study is to make sure that it is reproducible. Any researcher interested in doing so should find in your report or paper enough information to reproduce your study. This is why transparency in reporting a study is paramount. Let us first discuss what reproducibility means in the context of statistics.

Table 14.1: This table outlines the essential steps involved in any statistical project, where the primary objective is to analyse a sample of objects and infer key properties from collected data

Step	Description	Remarks
1	Write one or multiple research questions	A good research project starts **always** with one (or multiple) **research question** (RQ). We can loosely define it as a concise, focused inquiry formulated to address a specific concern or knowledge gap within a broader topic area
2	Design multiple hypotheses	Once you formulate your RQ, you will need **hypotheses**. Something you can disprove or verify. Hypotheses can be loosely defined as a prediction about the relationship between two or more variables. It can be described as an educated guess about what happens in an experiment
3	From hypotheses design **population eligibility criteria**, **sampling approach** (random, non-random, etc.), and **data collection strategy** (define budget, sample size, etc.)	This step is fundamental to be able to get the right data samples for your study. Remember that populations must be defined to be useful in answering your RQs and to support (or disprove) your hypotheses
4	Design the experiments	An experiment is a methodical procedure carried out with the objective of verifying or falsifying one or multiple hypotheses. Experiments involve manipulating one or more variables to determine their effect on a certain outcome
5	Collect data	Your main task is to get the data you need from your sample. This may involve surveys, measurements, etc.
6	Do the **analysis**	Do the analysis you planned. Describe first the data with measures of central tendency (see Chap. 4), dispersion (see Chap. 5), and position (see Chap. 6). Use visualisation to show your data and to better understand its distribution
7	If applicable do **hypothesis testing** (e.g. is group A different in some way than group B?).	Hypothesis testing in statistics is a method used to evaluate the validity of a claim about a population parameter by analysing sample data

14.2.1 Reproducibility

Reproducibility refers to the ability to duplicate the results of a study or experiment using the same data and statistical methods. It is a cornerstone of scientific research that ensures that the findings are reliable, credible, and **verifiable**. Reproducibility is achieved when independent researchers can obtain the same results following the methodology of the original study, using the same data, performing the same data processing and analysis.

14.2 Reproducibility, Replicability, Transparent Reporting, and Documentation

Warning 14.2.1: **Reproducibility and Replicability**

In research, we talk about reproducibility and replicability. They have different meanings, and both are important in research.

Reproducibility refers to the ability to duplicate the results of a study using the **same data** and **methods** as the original study. It means that independent researchers can take the original dataset and the exact analysis pipeline (including code, software, and statistical methods) to obtain the same results. It ensures that the data processing and analysis steps are transparently reported and can be followed by others, confirming the reliability of the computational aspects of the research.

Replicability refers to the ability to duplicate the results of a study by conducting a **new study** under the same experimental conditions. This involves collecting **new data** but following the same experimental protocol as the original study. The focus here is on verifying the findings by generating new data. It ensures that the scientific findings are robust and can be observed under the same conditions but with different samples, confirming the generalisability of the results.

There are several aspects that are needed to make a study reproducible.

- **Data availability**: The original data used in the study must be accessible to other researchers. This includes raw data, cleaned data, and any intermediate datasets generated during the analysis process. In Sect. 14.2.2 I list a few of the most known websites that allow you to share data.
- **Methodological transparency**: The methods and procedures used in the study must be thoroughly documented. This includes the design of experiments, sampling techniques (see Chap. 3), data processing steps, and statistical analyses.
- **Code and software**: Any code or software scripts used for data analysis should be available. This ensures that others can execute the same analysis steps, using the same parameters and algorithms. In Sect. 14.2.2 I list a few of the most known websites that allow you to share code.
- **Documentation**: Detailed documentation is crucial. This includes descriptions of the data and of the data collection strategy, explanations of the analysis steps, justifications for methodological choices, and notes on any assumptions made during the analysis.
- **Results verification**: The ability to replicate the results strengthens the validity of the original findings. Researchers should aim to provide all the necessary information so that others can independently verify the results.

Reproducibility is fundamental for various reasons. The main one is scientific integrity. Reproducibility allows for the verification of results and ensures that scientific conclusions are based on reliable evidence. Forcing researchers to make a study reproducible helps reduce fraud and dishonest practices. When results are reproducible, they inspire greater confidence among researchers, policymakers, and the public. This is essential for the application of scientific findings. Furthermore, reproducible research contributes to the cumulative nature of scientific knowledge. This allows new studies to build on previous work, advance the field, and foster innovation. Finally, reproducibility helps identify errors or inconsistencies in research. If results cannot be replicated, they may indicate issues with the original data, methods, or analyses.

Note that it is not always easy to ensure reproducibility for various reasons. For example, data privacy and confidentiality can make data sharing impossible. This, of course, can hinder reproducibility efforts, but there is not much one can do about it. This happens often when dealing with medical data, where privacy is a fundamental aspect of the ethics of medical studies. Sometimes infrastructure may play a role in reproducing a study. Maybe laboratory equipment is very expensive or more specialised training is required to operate some measurement instruments. Unfortunately, that is the price to pay when doing cutting-edge research on very complex topics. However, and maybe exactly for these reasons, it is paramount that you document every single small detail of your study, to make reproducibility **possible**, even when not easy.

You should always share the data and code of your study. Today, it is very easy to do so and will enable other researchers to validate and reproduce your results and as a by-product will force you to write better code.

14.2.2 Data and Code Sharing

Let me say this again: Your research should **always** be reproducible. To guarantee this, it is mandatory to share the data and code with everyone when publishing your results. In this way, your results can be verified and will gain credibility. Several web platforms exist to share data and code. Here are some of the most known at the time of writing.

- **Data sharing**: Figshare (https://figshare.com/), Zenodo (https://zenodo.org/), Dryad (https://datadryad.org/stash), and Mendeley Data (https://data.mendeley.com/).
- **Code sharing**: GitHub (https://www.github.com) or Bitbucket (https://bitbucket.org/). GitHub is a good choice, since it allows one creating a beautifully structured code repository with long and detailed *readme* files with instructions, links, references, and much more.

All of them are widely recognised by almost all scientific journals and are known by researchers. You cannot go wrong with any of those. When sharing the data, typically you get a **D**igital **O**bject **I**dentifier (**DOI**)[1] that you can use to share the dataset directly with other colleagues.

---

[1] A DOI, or Digital Object Identifier, is a unique alphanumeric string assigned to a digital object, such as a journal article, research paper, dataset, or any other piece of intellectual property that exists in a digital form. The DOI provides a permanent Internet link to the digital object, ensuring that it can always be found and accessed, even if the location of the object changes over time.

14.2.3 Transparent Reporting

To ensure reproducibility, you need to report **everything** about your study and give clear instructions about what you have done. This will allow others to verify your results and hopefully build on your work to further push research in interesting directions. Furthermore, detailed and transparent documentation holds researchers accountable for their methodologies and conclusions. It ensures that all steps of the research process are visible and open to scrutiny. This will make fraudulent and unethical studies difficult and increase trust in science and statistics. Finally, clear documentation facilitates dissemination of knowledge. It may allow researchers in different fields to grasp the significance of the study and potential applications. Here are some tips to make your reporting transparent.

- **Methodological details**:
  - **Study design**: Describe the study design in detail, including experimental or observational approaches, controls, and randomisation methods.
  - **Data collection**: Document how data were collected, including instruments used, measurement procedures, and any calibration details. If the data were collected by a team, explain how the team was chosen, how they approached the subjects (if relevant), etc. Describe how you have defined your population (eligibility criteria, see Chap. 3), and give exact information on it (means, ranges of values, etc.).
  - **Sample size and selection**: Explain how the sample size was determined and the criteria for including or excluding subjects from a study.
  - **Outliers**: Describe how you defined outliers (see Chap. 7) and whether you have removed them or not. If you remove data points, you should always explain why and present the same results with and without outliers, to make clear what is their effect on your conclusions.

- **Statistical Analysis**:
  - **Software and tools**: Specify the software and tools used for data analysis, including versions and any custom scripts or code. Again share your code. Generally, try to avoid tools that hide details of method implementations. Your study should explain what method you used, not which button you pressed in a tool. Your study must be reproducible with any statistical software package. If your tool does not give you information on how a specific method is implemented, change the tool.
  - **Data processing**: Describe the steps taken to process and clean the data, including any transformations or normalisation procedures.
  - **Analysis techniques**: Provide a comprehensive account of the statistical techniques employed, including any assumptions made and how they were tested. Everything we discuss in this book should be documented, explained, and justified when used.

- **Result presentation**:
  - **Descriptive statistics**: Report basic descriptive statistics, such as means, medians (see Chap. 4), variances, and standard deviations (see Chap. 5), along with appropriate visualisations (see Chap. 10).
  - **Inferential statistics**: Clearly present the results of hypothesis tests (see Chap. 12), hypotheses, confidence intervals (see Chap. 11), effect sizes, and any other inferential statistics used.
  - **Graphs and tables**: Ensure that all graphs and tables are well labelled and accompanied by explanatory captions that make the results easy to interpret. Recall that, whenever possible, it should be possible to understand the content of a figure from its caption (at least the main idea of it). Choose the right plot type to present your data (see Chap. 10).
- **Data sharing**:
  - **Raw data**: Where possible, share raw data or provide access to datasets through repositories or supplementary materials (see Sect. 14.2.2).
  - **Code and algorithms**: Share the code and algorithms used for data analysis to facilitate reproducibility and allow others to verify the findings (see Sect. 14.2.2). Avoid over engineering your code to make it understandable. That means avoid creating nestled functions, and make it as simple as it can possibly be.
- **Ethical considerations**:
  - **Ethical approval**: Report any ethical approvals obtained and describe how ethical guidelines were adhered to during the study. This is a key aspect in medical studies that goes beyond the scope of this book. If you work with medical data, you are probably in contact with a hospital. They surely have an ethical committee that you can contact for help.
  - **Conflicts of interest**: Disclose any potential conflicts of interest that could have influenced the research outcomes. Most journals ask authors to declare if and what conflict of interest they might have.

14.2.4 Best Practices for Documentation

It is important to also briefly discuss how to keep a good documentation. You will not remember what you did 2 weeks ago, let alone a few months ago. Professional record keeping is paramount for any scientific endeavour. Maintain thorough records of all research activities, including decisions made during the study, changes in methodology, and justifications for those changes. A notebook is a fundamental tool for any researchers. When I was studying physics, the rule was you had a bounded notebook with numbered pages. No page could be removed, and the numbers would prove if that would be the case. Negative results, errors, and strange findings need

14.2 Reproducibility, Replicability, Transparent Reporting, and Documentation

to be documented in your notebook. You never know where your next discovery is coming from!

Use version control systems to manage changes to documents, data, and code. This practice ensures that all modifications are tracked and that previous versions can be retrieved if necessary. GitHub, for example, is a great tool for this. Versioning data may be more difficult, especially if datasets are large. For example, at the moment of writing, GitHub has a file size limit of 100 Mb, making storing data often not possible. Publish research findings in **open-access journals**, and make data and materials freely available to the research community. Always write clearly and concisely, avoiding jargon and ensuring that the documentation is accessible to both specialists and, at least in some measure, to non-specialists.

Glossary

The following are terms that you are likely to encounter with a short description.

Alternative Hypothesis The hypothesis that there is a significant difference between groups or the expected relationship in a study

ANOVA (Analysis of Variance) A collection of statistical models used to analyze the differences among group means and their associated procedures

Average (Mean) The sum of a collection of numbers divided by the count of numbers in the collection

Bias The systematic error introduced into sampling or testing by selecting or encouraging one outcome or answer over others

Confidence Interval A range of values, derived from sample statistics, that is likely to contain the value of an unknown population parameter

Correlation A measure of the relationship between two variables and their dependence on one another

Hypothesis Testing A method of statistical inference to determine the probability that a hypothesis concerning a population parameter is true

Interquartile Range (IQR) The difference between the third quartile and the first quartile, representing the middle 50% of the data

Median The middle value in a dataset, which divides the set into two equal halves

Mode The most frequently occurring value in a dataset

Null Hypothesis A hypothesis that assumes no statistical significance exists in a set of given observations

Outlier An observation that lies an abnormal distance from other values in a random sample from a population

***p*-Value** The probability of observing test results at least as extreme as the results actually observed, under the assumption that the null hypothesis is correct

Percentile A measure indicating the value below which a given percentage of observations in a group of observations fall

Probability A measure of the likelihood that an event will occur

Quartile A type of percentile that divides the data into four defined intervals

Regression A statistical method for estimating the relationships among variables

Sample Size The number of observations or replicates included in a statistical sample

Statistical Significance The likelihood that a result or relationship is caused by something other than mere chance

Variance The average of the squared differences from the Mean, showing how spread out the data points are

***z*-Score** The number of standard deviations a data point is from the mean

Appendix A
★ Partitioning of the Ordinary Least Square Variance

All is well that ends well

This section outlines the proof that R^2 can be understood as the ratio between the variance that the fitted model can explain and the total variance of the data. Note that intermediate linear algebra knowledge is required to understand this section. We will work in matrix notation and will write the linear regression formula as

$$y = X\beta + \epsilon \tag{A.1}$$

Shape: y is $n \times 1$, X is $n \times k$, β is $k \times 1$, ϵ is $n \times 1$.

where y is a vector with all the measurements of the dependent variable. X is a matrix where each column of the matrix X is an observation or measurement (made of the independent variables) and with the peculiarity that the first column is made of all ones (to keep into account the constant factor in the linear formula), and β is a vector containing the coefficients (which in our previous example have denoted by α_i). ϵ is a vector that contains the remaining errors (the part that linear regression cannot explain). ϵ is also called the **residual** between the prediction ($X\beta$) and the true value y.

It is known that the optimal coefficients of linear regression $\hat{\beta}$ can be written in closed form as [16]

$$\hat{\beta} = (X^T X)^{-1} X^T y \tag{A.2}$$

We can now write the total variance considering that the residual $\hat{\epsilon}$ is given by (see Eq. (A.1) in the case of the optimal parameters $\hat{\beta}$) $y - X\hat{\beta}$. Furthermore, we can write the residual sum of squares (RSS) of the residuals, which is given by $\hat{\epsilon}^T \hat{\epsilon}$ and from the previous equations

$$\text{RSS} = y^T y - y^T (X^T X)^{-1} X^T y \tag{A.3}$$

Denote by $\bar{y}$ a vector of dimension $n \times 1$ with elements equal to the average y_a of the elements of y. In other words

$$\bar{y} = \begin{pmatrix} y_a \\ y_a \\ \vdots \\ y_a \end{pmatrix} \tag{A.4}$$

The total sum of squares (TSS) is given by

$$\text{TSS} = (y - \bar{y})^T (y - \bar{y}) = y^T y - 2 y^T \bar{y} + \bar{y}^T \bar{y} \tag{A.5}$$

The explained sum of squares (ESS), defined as the sum of squared deviations of the predicted values from the observed mean of y, is given by

$$\text{ESS} = (\hat{y} - \bar{y})^T (\hat{y} - \bar{y}) = \hat{y}^T \hat{y} - 2 \hat{y}^T \bar{y} + \bar{y}^T \bar{y} \tag{A.6}$$

Now using $\hat{y} = X\beta$ and simplifying, it can be shown that

$$\text{TSS} = \text{ESS} + \text{RSS} \tag{A.7}$$

if and only if $y^T \bar{y} = \hat{y}^T \bar{y}$. Since $\bar{y} = (y_a, y_a, \ldots, y_a)$ (where y_a is the scalar average of the values contained in y), this condition is simply

$$y_a \sum_{i=1}^n y_i = y_a \sum_{i=1}^n \hat{y}_i \tag{A.8}$$

and simplifying by dividing by y_a

$$\sum_{i=1}^n y_i = \sum_{i=1}^n \hat{y}_i \tag{A.9}$$

in words TSS = ESS + RSS is true if and only if the sum of the predictions is equal to the sum of the expected values (or that the sum of residuals is zero). That this is true can be shown as follows. Consider $X^T \hat{\epsilon}$. It can be shown that $X^T \hat{\epsilon} = 0$. In fact

$$\begin{aligned} X^T \hat{\epsilon} &= X^T \left[I - X(X^T X)^{-1} X^T \right] y \\ &= (X^T - X^T X (X^T X)^{-1} X^T) y = 0 \end{aligned} \tag{A.10}$$

since clearly $X^T X (X^T X)^{-1} = I$ with I the identity matrix. At the same time

$$X^T \hat{\epsilon} = \begin{pmatrix} 1 & 1 & \ldots & 1 \\ X_1^{[1]} & X_1^{[2]} & \ldots & X_1^{[n]} \\ \ldots & \ldots & \ldots & \ldots \\ X_k^{[1]} & X_k^{[2]} & \ldots & X_k^{[n]} \end{pmatrix} \begin{pmatrix} \hat{\epsilon}_1 \\ \hat{\epsilon}_2 \\ \ldots \\ \hat{\epsilon}_n \end{pmatrix} \tag{A.11}$$

Now we do not need to perform the entire multiplication. It is enough to note that the first row of $X^T \hat{\epsilon}$ from the last equation equals the sum of the $\hat{\epsilon}_i$, but we know from Eq. (A.10) that this $X^T \hat{\epsilon} = 0$; thus also the first row (the sum of the residuals) must be equal to zero. This proves our statement, and we can conclude that TSS = ESS + RSS. Note that this is only true in this case of linear regression and is not valid in general. So be very careful in using R^2 in a case where you do not know what kind of relationship you have. Its interpretation is tricky, and the fact that TSS = ESS + RSS is technically only valid in the case of pure linear regression.

Appendix B
Big-*O* and Little-*o* Notation

B.1 Big-*O* Notation

Big-*O* notation is a mathematical concept used to describe how fast a function grows when its argument goes to a certain value or infinity. In computer science often it is used to measure the efficiency of an algorithm, particularly in terms of its time or space complexity as the input size grows. In this case, it provides an upper bound on the growth rate of an algorithm's runtime or memory usage (and, in general, of a function), helping to understand the worst-case scenario of its performance.

Imagine you have a task, like sorting a list of numbers. As the list gets longer, the time it takes to sort the list usually increases. Big-*O* notation helps us to understand how fast the sorting time increases as the list length grows. In more formal terms, consider a function $f(x)$ and $g(x)$ a positive function for large values of x. The formula

$$f(x) = O(g(x)) \tag{B.1}$$

is read as follows: $f(x)$ is big-*O* of $g(x)$ if there is an $M > 0$ and a real number x_0 such that

$$|f(x)| \leq M g(x) \quad \forall x > x_0 \tag{B.2}$$

Imagine now that we are evaluating the time needed by an algorithm to finish as a function of the length n of the input. The Big-*O* notation allows us to classify algorithms by their efficiency. Here are some Big-*O* examples.

- **O(1)**: Constant time. The algorithm's runtime does not change with the input size.
- **O(log n)**: Logarithmic time. The runtime grows logarithmically as the input size increases. Or better phrased, the runtime **does not** grow **faster** as a $\log n$ function.
- **O(n)**: Linear time. The runtime grows linearly with the input size.

B.2 Little-*o* Notation

Intuitively, the statement $f(x)$ is $o(g(x))$ (read as $f(x)$ is little-*o* of $g(x)$) means that $g(x)$ grows much faster than $f(x)$. In more formal terms, consider a function $f(x)$ and $g(x)$ a positive function for large values of x. The formula

$$f(x) = o(g(x)) \tag{B.3}$$

means **for every** $\epsilon > 0$ and a real number x_0 it is true that

$$|f(x)| \leq Mg(x) \ \forall x > x_0 \tag{B.4}$$

Note that this must be valid **for all** ϵ values. In the Big-*O* notation case, Eq. (B.2) should only be valid for at least one M.

References

1. Robert V Hogg, Elliot A Tanis, and Dale L Zimmerman. *Probability and statistical inference*, volume 993. Macmillan New York, New York, 1977.
2. George Casella and Roger L Berger. *Statistical inference*. Cengage Learning, 2021.
3. Christopher Engledowl and Travis Weiland. Data (Mis)representation and COVID-19: Leveraging Misleading Data Visualizations For Developing Statistical Literacy Across Grades 6–16. *Journal of Statistics and Data Science Education*, 29(2):160–164, May 2021.
4. States accused of fudging or bungling COVID-19 testing data, May 2020.
5. Spurious Correlations. https://www.tylervigen.com/spurious-correlations (last accessed 4th May 2024).
6. Giacomo Livan, Marcel Novaes, and Pierpaolo Vivo. *Introduction to Random Matrices - Theory and Practice*, volume 26. 2018. arXiv:1712.07903 [cond-mat, physics:math-ph].
7. Rensis Likert. A technique for the measurement of attitudes. *Archives of psychology*, 1932.
8. Changbao Wu, Mary E Thompson, Geoffrey T Fong, Qiang Li, Yuan Jiang, Yan Yang, and Guoze Feng. Methods of the International Tobacco Control (ITC) China survey. *Tobacco Control*, 19(Suppl 2):i1–i5, 2010.
9. Bradley Efron. Bootstrap methods: another look at the jackknife. In *Breakthroughs in statistics: Methodology and distribution*, pages 569–593. Springer, 1992.
10. Michael R Chernick. *Bootstrap methods: A guide for practitioners and researchers*. John Wiley & Sons, 2011.
11. A. C. Davison and D. V. Hinkley. *Bootstrap Methods and their Application*. Cambridge Series in Statistical and Probabilistic Mathematics. Cambridge University Press, Cambridge, 1997.
12. Justin Matejka and George Fitzmaurice. Same Stats, Different Graphs: Generating Datasets with Varied Appearance and Identical Statistics through Simulated Annealing. In *Proceedings of the 2017 CHI Conference on Human Factors*

in Computing Systems, pages 1290–1294, Denver Colorado USA, May 2017. ACM.
13. Patrick Wessa. Free statistics software. *Office for research development and education, version*, 1(1.2018), 2012.
14. Rob J. Hyndman and Yanan Fan. Sample Quantiles in Statistical Packages. *The American Statistician*, 50(4):361–365, 1996. Publisher: [American Statistical Association, Taylor & Francis, Ltd.].
15. Stephen Abbott. *Understanding Analysis*. Undergraduate Texts in Mathematics. Springer, New York, NY, 2015.
16. Umberto Michelucci. *Mathematical Concepts for Machine Learning in Science*. Springer Nature, New York, 2024.
17. Journal of Statistics Education, v13n2: Paul T. von Hippel, February 2016.
18. Arthur Lyon Bowley. *Elements of statistics*. Number 8. King, 1926.
19. George Udny Yule. *An introduction to the theory of statistics*. C. Griffin, 1927.
20. Peter G Bryant and Marlene A Smith. Practical data analysis: Case studies in business statistics, Richard d, 1995.
21. Irving W Burr. Cumulative frequency functions. *The Annals of mathematical statistics*, 13(2):215–232, 1942.
22. Antony Unwin and Kim Kleinman. The iris data set: In search of the source of virginica. *Significance*, 18(6):26–29, 2021.
23. C. Spearman. The Proof and Measurement of Association between Two Things. *The American Journal of Psychology*, 15(1):72–101, 1904. Publisher: University of Illinois Press.
24. M.G. Kendall. New measure of rank correlation. *Biometrika*, 30(1-2):81–93, 1938.
25. Trevor Hastie, Robert Tibshirani, James Gareth, Daniela Witten, and Jonathan Tylor. *An Introduction to Statistical Learning - with Applications in Python*. Springer Nature.
26. Trevor Hastie, Jerome Friedman, and Robert Tibshirani. *The Elements of Statistical Learning*. Springer Series in Statistics. Springer, New York, NY, 2001.

Index

A
Analysis
   data, 14

B
Bias
   data sampling, 5
Bitbucket, 150
Bootstrap, 34, 126

C
Causality, 6
Causation and correlation, 5
Central tendency
   measures, 39
Christopher Engledowl, 3
Clustering, 30
Code
   sharing, 150
Coefficient
   Pearson mode skewness, 99
Cohort, 18
Confidence interval
   mean difference, 125
   proportions, 125
Conflict of interest, 152
Correlation, 137
   spurious, 5
Correlation and causation, 5
Countable, 9
Criteria
   eligibility, 27
Cumulative distribution function (CDF), 77

D
Dangers
   relying on single statistics, 52
Data
   binary, 20
   categorical, 15
   categorical, definition, 15
   categorical, nominal, 15
   categorical, ordinal, 15
   collection, 23, 151
   continuous, 17
   cross-sectional, 20
   dichotomous, 20
   discrete, 17
   longitudinal, 19
   norminal, 15
   ordinal, 15
   qualitative, 15
   qualitative, definition, 15
   quantitative, 16
   quantitative, definition, 17
   raw, 152
   sharing, 150
Data analysis, 14
Data collection
   bias, 6
Datasaurus
   dataset, 52
Decile, 64
   definition, 65
Descriptive statistics, 12
Dice, 8
   six-face, 9
Digital Object Identifier (DOI), 150
Distribution

bimodal, 104
binomial, 88
binomial, PMF, 88
continuous, 74, 75
discrete, 74
heavy tailed, 96
light tailed, 96
mean, 79
multimodal, 104
normal, 78
normal, PDF, 79
Poisson, 90
standard deviation, 79
uniform, 104
unimodal, 104
Documentation, 149
Dryad, 150

E
Eligibility criteria, 27
Ethical approval, 152
Event, 8
definition, 9
Expected value, 40, 77
Experiment
random, 8

F
Figshare, 150
Function
moment generating, 105

G
GitHub, 150

H
Hypothesis, 24, 25, 148
Hypothesis testing, 129

I
Inferential statistics, 13
Interquartile range (IQR), 63, 64
definition, 64

K
Karl Popper, 25
Kurtosis, 102

L
Linear regression, 141
Ludwig Wittgenstein, 8

M
Mean, 39

definition, 40
example, 41
Measure
position, 55
Measurement
level, interval, 18
level, nominal, 17
level, ordinal, 17
level, ratio, 18
Median, 39, 42
definition, 42
example, 43
Mendeley Data, 150
Methodological transparency, 149
Mid-range, 39, 44
definition, 45
example, 45
Modality, 103
Mode, 39, 43, 44
definition, 44
example, 44
Moment generating function (mgf), 105

O
Open-access journals, 153
Outcome space, 8
definition, 8
Outlier, 67, 151
causes, 70
domain-specific criteria, 68
impact, 70
IQR method, 67, 68
treatment, 70
z-score method, 68

P
Pearson coefficient, 137
Pearson median skewness coefficient, 100
Pearson mode skewness coefficient, 99
Percentile, 56
definition, 62
Poisson distribution, 90
Poisson process
approximate, 89
Population, 2
Probability mass function (PMF), 74
six-face dice, 75
p-values, 134

Q
Quartile, 62
definition, 62
first, 62
second, 62
third, 62

R

Random experiments, 8
    definition, 8
Random sampling without replacement, 31
Random sampling with replacement, 32
Random stratified sampling, 33
Random variable, 9, 10
Range, 51
    definition, 51
Replicability, 149
Reporting
    transparent, 151
Reproducibility, 148
    lack of, 6
Research question, 24, 148

S

Sample, 2
Sampling
    convenience, 28
    non-probability, 28
    non-probability, process, 29
    probability, 29
    probability, process, 30
    quota, 28
    random, stratified, 33
    random without replacement, 31
    random with replacement, 32
    restricted, 28
    stratified, 33
    stratified, process, 33
Sampling theory, 23
Skewness
    Pearson median coefficient, 100
    Pearson mode coefficient, 99
Space
    outcome, 8
S^2 sample variance, 48
Standard deviation, 49
    definition, 50
Statistics
    bad, 3
    cherry picking data, 5
    descriptive, 12, 152
    descriptive, example, 12
    inferential, 13, 152
    inferential, example, 13
    misleading presentation, 5
Stratification, 30
Study design, 151
Survey
    meaning, 27
    sampling, 26, 27

T

Tails
    heavy, 96
    light, 96
Travis Weiland, 3
Tyler Vigen, 5
Type I error, 135
Type II error, 135
Types of data, 15

V

Variable
    continuous, 9
    random, 9
    random, countable definition, 9
    random, definition, 10
    random, discrete, 9
Variance, 47, 77
    definition, 47
    population, 48
    sample, 48

W

Winston Churchill, 5

Z

Zenodo, 150
z-score
    example, 69
    method, 68

The manufacturer's authorised representative in the EU is Springer Nature Customer Service Centre GmbH, Europaplatz 3, 69115 Heidelberg, Germany. If you have any concerns regarding our products, please contact ProductSafety@springernature.com

Printed and bound by CPI Group (UK) Ltd, Croydon, CR0 4YY
05/03/2026
02065353-0001